> Dear Todd & Kee
> I did it! I'm sure you
> can find some good
> info in here — ignore
> the doomsday tone
> and the 8th grade
> mentality of the writing
> — S. Eubank

How to Protect Your
Life & Property

How to Protect Your Life & Property

By

MILO A. SPERIGLIO

and

S. THOMAS EUBANKS

Illustrated by Randy J. Leifer
Cover Design by Todd Waite

SEVILLE PUBLISHING
SHERMAN OAKS, CALIFORNIA

Copyright © 1982 by Milo A. Speriglio
and S. Thomas Eubanks

All rights reserved. Except for brief use in
a review, the reproduction or utilization of this work
in any form or by any electronic, mechanical, or other
means, now known or hereafter invented, including xerography,
photocopying, and recording, and in any information storage
and retrieval system is forbidden without the written permission
of the publisher and the authors.

Library of Congress Catalog Card No. 81-52905

ISBN 0-930990-01-3

Manufactured in the United States of America

SEVILLE PUBLISHING COMPANY

14410 Ventura Boulevard
Sherman Oaks, Ca 91423
(213) 501-5200

PREFACE

We thank the entire staff of Nick Harris Detectives, Inc. for their assistance in helping assemble this book, and use of their vast resources.

Without the cooperation of law enforcement bureaus throughout the nation, this book could not be possible, so we offer many special thanks to the hundreds of policemen and women who shared with us their expertise, as well as the assistance from numerous government agencies who gave us access to vital information.

To present our readers with the most comprehensive view of the subject matter, we had to abstract sensitive information from victims of crimes, as well as the criminals who committed them.

In order to gain this incredible information, we guaranteed immunity to the victims, their families, informants, and admitted criminals who openly told us their stories.

Many of the perpetrators of violent crimes who consented to tell us their "trade secrets" did so to help you protect your life and property. These persons include countless infamous criminals of modern time.

Most of them are presently serving time in prison. Others are now on probation or parole. A few have reformed and have not spent as much as one day in jail.

These persons have murdered, mugged, robbed, and raped innocent victims. They revealed how they committed their crimes so we could report to you how you can avoid becoming a victim.

There are probably a thousand names to thank for contributions made to this book. The most important thank you, though, is that this book will save many lives—maybe yours—and for that we are grateful to all those who gave us their time, experience, and compassion so that you could learn, *"How To Protect Your Life & Property."*

M.A.S. & S.T.E.
Los Angeles, CA
January, 1982

About the Authors

MILO A. SPERIGLIO

Internationally-known professional investigator and criminologist for over two decades. Speriglio is the Director and Chief of Nick Harris Detectives, Inc., founded in 1906 and headquartered in Los Angeles, California.

In 1977 he was a candidate for Mayor of Los Angeles, and challenged Mayor Tom Bradley and tax crusader Howard (Proposition 13) Jarvis. The *Los Angeles Times* reported: "Speriglio has come out with the most sweeping program of the dozen mayoral candidates to combat crime."

Co-author Speriglio, is frequently on local and network radio and television. He is often read about in newspapers and magazines.

He is listed in *Who's Who in America—West, Who's Who in American Law Enforcement* and *Who's Who in California*. Among Speriglio's most publicized investigations was the death of Marilyn Monroe. His findings cast doubt upon her alleged suicide, indicating that it was political assassination.

Speriglio is certified by the California Department of Justice to license citizens to carry tear gas for self-protection. He has trained thousands of persons in the art of basic self-defense, use of tear gas weapons, and how to protect their lives and property.

He is the co-author of *WOMEN... THE DELICATE PREY* (with Boyd Upchurch), an awareness and preparedness book about rape, soon to be released by Seville Publishing Company.

S. THOMAS EUBANKS

Following a lucrative career as a full-time writer and actor, in 1973 Eubanks merged his bellitristic abilities with private investigation. Since then he has written numerous training manuals and investigative training films as the Director of the Nick Harris Detective Academy, founded in 1911.

Since 1978 he has directed operations as the Executive Director and Vice-President of Nick Harris Detectives, Inc.

Eubanks will soon have published a novel, *JOE BELL,* and a collection of short stories tentatively titled, *PIGEON-HOLED & FORGOTTEN.* He is the co-author of a screenplay called *MOONDOGS,* and another still rolled around the platten, *GUN METAL GRAY CHEVROLET.*

DEDICATION

May their lives be long and protected:

Patricia, my wife, and Holly & Janelle, my daughters.
To my mother, Sylvia, and sister Amelia.
Co-author Milo A. Speriglio

Judy, my wife, and Cassandra, my daughter.
Co-author S. Thomas Eubanks

We also dedicate our book to all our readers, and hope their lives will also be long and protected.

TABLE OF CONTENTS

INTRODUCTION

Paranoia: Are We Overreacting? xvii

National Crime Clock xvii

Definition of Crime xx

    Homicide
    Aggravated Assault
    Forcible Rape
    Robbery
    Burglary
    Larceny-Theft
    Motor Vehicle Theft
    Simple Assault
    Arson
    Forgery & Counterfeiting
    Fraud
    Embezzlement
    Buying, Receiving, or Possessing Stolen Property
    Vandalism
    Prostitution
    Narcotic Drug Laws
    Gambling
    Disorderly Conduct

Murder Victims: The Powerful,
The Famous, and The Unknown xxv

BOOK I

HOW TO PROTECT YOUR LIFE

CHAPTER I—VIOLENT CRIMES

The Last Day of Your Life 29
Stick 'Em Up 33
Rape: When, Where, What and Why 41
Approach Your Car With Caution...................... 45
Myths, Misunderstandings, and Categories of Rape..... 49
Submit or Resist? 50
The 40 Most Dangerous Cities for Females............. 51
Rape Crisis Hotline Numbers........................... 53
How Safe Are Universities and Colleges? 58
A Spotlight For The Bad Guy 60
Mr. D.A. 61
Criminals Kill Police Officers Too 61
Your Child Could Be Molested 63
Compensation For Victims of Violent Crimes........... 66
"WE TIP Hotline" For the Anonymous Caller 68

CHAPTER II—EVERYDAY SURVIVAL TIPS

The Obscene Phone Call 73
Dangerous and Deadly Toys 75
Don't Advertise You Live Alone 76
Growl 77
Safety At Home For Seniors 78
Special Warning For Senior Citizens 79
When Have You Had Too Much To Drink To Drive?..... 80

*How Do You Protect Yourself
   From "The Drunk Driver"?* 83

Does Your Youngster "Do Drugs"? 85

**CHAPTER III—DISASTER, EMERGENCY
           & MEDICAL PROTECTION**

Earthquakes 89

Flood 91

Lightning 91

House Fires 92

What You Need To Know About Fire Extinguishers 93

How To Deal With Fires 94

    Kitchen Grease Fires
    Electrical Fires
    Fire Drills

How To Survive a Hotel or Apartment Fire 96

Household Emergencies & Safety Precautions 98

    Power Failure
    Appliance Smoking or Burning
    Downed Power Line
    Plumbing Emergencies
    Appliance Shock
    Trash Compactors
    Small Engines
    Chain Saws
    Lawn Mowers
    Pesticides
    Household Insect Control
    Power Tools
    Air Conditioning Emergencies
    Heating Emergencies

Physicians: Your Life Is In Their Hands 110

Medications: How To Save Money & Your Life 111

CHAPTER IV—EVERYDAY SELF-DEFENSE

Common Self-Defense Weapons 115

    Attack From Behind
    Attack From the Front
    The Voice
    Self-Defense Against Rapists

Tear Gas: Answers To All Your Questions 123

    Are You Prepared?
    The Controversy
    Everything You Need To Know About Tear Gas
    Your Only Protection Is Self-Protection

The Pros & Cons of Handguns 148

BOOK II

CONSUMER PROTECTION

CHAPTER I—SCHEMES & SCAMS

Interview With A Con Man............................ 153

    Short-Changing
    Insurance Fraud
    False Identification
    The Money Game
    "Making" Money

Fraud: The Con In Confidence 160

Ripped Off Renters 162

Energy Scams... 165

Mail Order Robbery 166

Beware of Free or Cheap Government Land 169

Fix-It Fraud 170

The Great Pretenders................................. 171

CHAPTER II—PAPER, PLASTIC AND POVERTY

Cash to Ashes 173
Warning Signs of Bankruptcy 174
How To Save Money With Your Credit Card 175

CHAPTER III—CONTRACTS & WARRANTIES

Warranties: Read Before You Buy 177
Contracts: A Binding Legal Agreement 181
The Cooling Off Law 182

CHAPTER IV—PRODUCTS & SERVICES

Shopping Myths 185
    Paying For The Label
    Discounts and Sales
Buying Clothes 188
The Five Grades of Meat 190
Mirror, Mirror On The Wall 191
Solid Gold 192
How Safe Is Your Safe? 193
Life Insurance: Investment or Protection? 196
The Certainty Of Life 198

BOOK III
HOW TO PROTECT YOUR PROPERTY

CHAPTER I—BURGLARY

Introduction to Burglary 203
Every Second Counts 204
Record Your Valuable Property 208
Have You Set Out A Welcome Mat? 211
The Burglar's Thoughts 216

Keys Are Money 217

Routines & Answering Machines 221

CHAPTER II—PROTECTION ON VACATION

Vacation Security Preparation 223

Your Airline Reservation
Doesn't Guarantee You A Seat 225

Missing Luggage Syndrome 227

Traveler's Intruder Alarms 228

Crime Against Tourists: Vacationing In Paradise 229

CHAPTER III—AUTO THEFT

Abracadabra: There Go Your Wheels 243

   How They Get In
   How They Make A Stolen Car "Legal"
   When, Where & The Prevention
   Why They Steal Your Car & How They Get Rid Of It

Buying Or Selling A Car 260

How To Protect Your Car While In Repair 261

CHAPTER IV—PURSE-SNATCHING,
   PICKPOCKETING & PARENTAL KIDNAPPING

Purse-Snatching Prevention 265

Easy Pickin's For The Pickpocket 266

Parental Kidnapping 269

CHAPTER V—FORGERY, COUNTERFEITING,
   SHARKING & INDUSTRIAL ESPIONAGE

How To Write Your Checks 275

Your Money Cards 278

The Loan Sharks 281

How They Steal Your Company's Secrets 282

CRIME: THE SOLUTION 287

INTRODUCTION

Paranoia: Are We Overreacting?

Someday, this book may save your life.

Facts: crime in the United States rises nine times faster than the population growth. Each year a violent crime is committed against one member of every eight families. One out of every three females will be raped or assaulted sometime in their lifetime.

National Crime Clock

Every day someone is:

RAPED every 7 minutes.

MURDERED every 26 minutes.

ROBBED every 60 seconds.

BURGLARIZED every 10 seconds.

The "crime clock" ticks: every 29 seconds someone's car is stolen. A property crime is committed every 3 seconds; a violent crime every 27 seconds. During the period of time you read this book, nine persons will be murdered, forty raped, and 1,890 burglaries and robberies will occur.

Today we deadbolt, chain, construct peep holes, install alarms and take other security measures to safe-guard our lives and property. There was a time in our society when it was commonplace to leave the house unlocked. This practice has declined in even the smallest communities where everyone knows everyone.

Nationwide studies reveal that 80% of all persons will not open their door, unless they know the caller. Seventy-five percent of the population drive their cars with the doors and windows locked. Gun sales have reached a record high.

Are we overreacting?

Concerned citizens are now arming themselves with pocket and purse size tear gas weapons for self-protection.

Are we paranoid?

Crime—or even the threat of victimization—has frightened most of the nation. Many persons won't walk in the streets or drive at night. If you live in a large city, the odds are greater that you'll be murdered than be killed in a traffic accident.

The following are the most recent national annual crime statistics as reported in the FBI Uniform Crime Report. Two items of importance: 1) only reported crimes are listed (many crimes go unreported), and 2) each statistic represents a fellow human being:

Murder & Non-Negligent Manslaughter
21,456 persons murdered
An increase of 9.7%

Forcible Rape
75,989 rapes
An increase of 13.2%

Robbery
466,881 robberies
An increase of 12%

Aggravated Assault
614,213 assaults
An increase of 10.1%

Burglary
3,299,484 burglaries
An increase of 6.3%

Larceny-Theft
6,577,518 crimes
An increase of 9.9%

Motor Vehicle Theft
1,097,189 crimes
An increase of 10.6%

    Crime is just around the corner, ready to strike. Are you prepared to protect your life and property? Most law enforcement agencies agree: they can rarely be present to stop a violent crime from happening; your only protection is SELF-PROTECTION. Whatever you can do to prevent a crime or an attack from occurring will have to happen *before* the police arrive.

    Many crimes could have been prevented, if the victim had been informed, prepared. The purpose of this book is to frighten and prepare you to meet your attacker *before* he strikes. Within these pages is an education on how to avoid being attacked, how to protect your property and even how to sidestep the common "strings" used by rip-off artists. And you'll learn much more as this will be your guide to everyday survival.

    This is your book. Unlike anything you've ever read, it's about YOU. YOU are the protagonist. It concerns YOUR WELFARE. With over 41 MILLION crimes committed every year in America, you need to be concerned about your life, your property, and general survival in everyday living.

    Criminals know the odds are in their favor. They know they can escape arrest and punishment. This is the main reason America's crime reaches epidemic proportions.

Nature of the Crime	Unsolved Crimes
Murder	27%
Aggravated Assault	41%
Forcible Rape	52%
Robbery	75%
Burglary	85%
Larceny-Theft	81%
Motor Vehicle Theft	86%

Definition of Crime

To familiarize you with various types of crimes, we've defined them and included some important statistical information, of which most persons are unaware.

<u>Criminal Homicide</u>

Murder and non-negligent manslaughter. This is the willful and felonious taking of a human life.

Weather apparently has influence on murder. More murders occurred in the month of July.

Most frequently used weapons:

- 50% Handgun
- 4% Rifle
- 8% Shotgun
- 19% Cutting or stabbing
- 12% Other actual weapons or poison
- 7% Personal weapons, hands, feet, etc.

Approximately three out of four of the victims were male. 50% were white, 48% black, and 2% other races. Nearly one third of the victims were aged 20 through 29.

The average age of the offender is 18 to 22.

An almost even ratio of whites murdering whites and blacks murdering blacks were found.

About 1% of those murdered were infants. 9% were age 15 to 19, 33% aged 20 to 29, and 12% senior citizens.

Aggravated Assault

An unlawful attack to inflict bodily injury.

Like murder, this attack was more frequent in the month of July. Twenty-three percent of serious assaults were made with firearms. Knives or other cutting instruments were used in 22% of the assaults. Of all assaults, though, the hands, fists, and/or feet were most often used.

Forcible Rape

The carnal knowledge of a female through the use of force or the threat of force.

More rapes occurred in the month of August. Additional statistical information regarding rape is contained in several areas of this book.

Robbery and Burglary are the two most often confused crimes by the average person. "When I arrived home, I found everything gone; I'd been robbed." Wrong. As we'll see in the two following definitions, you were burglarized. Regardless of the name of the crime, though, you were a *victim*.

Robbery

Is the stealing or taking of anything of value from the care, custody, or control of a person, *in his presence,* by force or threat of force.

Almost a half of a million robberies are committed each year in the United States. If you ask most persons, particularly a store owner, what month of the year more robberies take place, they probably would say December. Retail stores, because of Christmas, do a high volume of sales, and have more cash in the till during that month. The most frequent month for robberies is not December, however, it's January.

The days of Jessie James are gone—or are they? Robbing a bank is easy. Most banks will not resist; they'll give you the money. Yet the average haul is only around $3,000. Not worth the risk of being caught. In addition to the local police investigation, the FBI investigates all bank robberies.

Gas station robberies and chain store robberies have drastically increased, but the highest increase in the past five years were among bank robberies, which also carries a higher risk of being caught and a sentence that is usually stiff.

Around 11% of all robberies took place in a private residence, and nearly half of them occurred in the street.

A firearm was used most frequently, and the age group of the offender in most robberies was 15 to 19, and from a standpoint of race, 57% of those arrested were black, 41% were white and 2% were other races.

Burglary

This is an unlawful entry of a structure to commit a felony or theft. For statistical purposes, this includes forcible entry, unlawful entry where no force was used, and attempted forcible entry.

Once again, a summer month, July, is the most frequent time burglary is committed. The age range of most offenders is lower than other crimes: 13 to 17 years of age. And whites scored 60%, while blacks accounted for 29%, and other races the remainder.

One reason burglary is expounded upon throughout this book is because about 20% of all crimes committed are burglary, resulting in well over three million reported crimes each year, and losses totalling over $2 billion.

Almost two-thirds of all burglaries are residential. Both residence and non-residence burglaries have sharply increased during the day-time hours.

Only 6% of all burglaries were committed by females. Persons aged 25 or less committed nearly 85% of the crimes.

Larceny-Theft

The stealing of property or articles *without* the use of force, violence or fraud. Purse-snatching, shoplifting, pick-pocketing, thefts from motor vehicles, and bicycle thefts are among the most common offenses. This crime doesn't include embezzlement, con-games, forgery or bad checks.

The dollar amount of loss, on the average, is just over $250. But it's still big business when you consider each year there are an average of 6,600,000 larcenous offenses.

No, we didn't forget to tell you in which month this crime most frequently occurs: again, it's August.

Females commit less crimes overall than males, but Larceny-Theft crimes attract more females than other crimes. Although they commit nearly 1/3 of them, more females are actually arrested for this crime. Shoplifting is one of their most frequent choices, and like burglary, the most frequent age group is 13 to 17.

Motor Vehicle Theft

The unlawful taking or stealing of a motor vehicle for any purpose including "joy riding."

On an average year about one million vehicles are stolen. When does this crime most often occur? You guessed it, August. Most persons have their vehicle insured, but when it's stolen, they usually receive only the wholesale value. On a typical $6,000 used car, at retail value, the wholesale is about $4,800.

Who's the most frequent violater of vehicle theft? A male between the ages of 13 to 17. Two-thirds of all offenders were white, and the rest black and other races.

We've provided you with statistical information and definitions of the most frequent types of crimes. Here are other crimes you should be aware of:

Simple Assault

Not an aggravated assault by nature.

Arson
Willful and malicious burning with or without intent to defraud, including attempted arson.

Forgery and Counterfeiting
Making, altering, uttering or possessing, with intent to defraud, anything (any object) which is false and made to appear true. This includes attempts.

Fraud
Fraudulent conversion and obtaining of money or property by false pretenses. This can include bad checks (not NSF, unless they knew there was no money in the account).

Embezzlement
Misappropriation of money or property entrusted to someone's care, custody or control.

Buying, Receiving or Possessing Stolen Property
This crime is defined by the title, and includes attempts.

Vandalism
Willful or malicious destruction, injury, disfigurement, or defacement of property (of any type) without the consent of the owner or person having custody or control.

Prostitution
Considered a non-violent crime. The offering of one's body (and/or services) for any consideration. This applies to both women and men.

Narcotic Drug Laws
Unlawful possession, sale, use, growing and manufacturing narcotic drugs.

Gambling

Another non-violent crime, committed by more Americans than any other crime. At one time in almost everyone's life, this law is violated.

Disorderly Conduct

Breach of peace.

There are thousands of other crimes in the law books of cities, counties, state and federal governments. They, too, are violated, and often, *you* become the victim.

Murder Victims:
The Powerful, The Famous, The Unknown

Roll Call: Victims Of American Violence During The Pitiless Decades Of The 60's, 70's And Now The 80's.
President John F. Kennedy, November 22, 1963
Dr. Martin Luther King, Jr., April 4, 1968
Senator Robert F. Kennedy, June 5, 1968
Beatle Star, John Lennon, December 8, 1980
Egyptian President, Anwar Sadat, October 6, 1981

They were all assassinated. Some other victims were luckier. Governor George C. Wallace was crippled by gunfire on March 15, 1972. President Ronald Reagan, his Press Secretary, James S. Brady, and others were wounded by gunfire during an assassination attempt on March 30, 1981, and Pope John Paul II was seriously wounded by a gunman on May 14, 1981.

They are household names for most persons. From the murder of President Kennedy eighteen years ago, to the attempted assassination of President Reagan, a total of 295,663 others were murdered in America.

Yes, you read it right, two hundred, ninety-five thousand, six hundred and sixty-three men, women and children were *murdered*. This is hard for anyone to grasp, but it is a fact, an unpleasant one.

With rare exception, most murder victims were totally unknown to the public, but at least ten million persons—friends and family of the victims—were saddened to learn of their deaths. But the other two hundred million citizens of the United States didn't know about their murder. Unlike the powerful and the famous, their tragedy wasn't captured by the TV news, and didn't receive media headlines.

How many of the past victims of murder could have survived their ordeal if they'd read this book? Not all, but many. Being in the wrong place at the wrong time can be dangerous. If and when you're confronted with danger, how you react or don't act can make the difference between life and death.

How well you're prepared is extremely important. We can't bring back those quarter of a million murdered persons or squelch their families' grief, but we can prepare *you* for everyday survival. And that's *exactly* what we intend to do.

BOOK ONE
HOW TO PROTECT YOUR LIFE

I
VIOLENT CRIMES

The Last Day Of Your Life

When you awoke this morning, you didn't expect today to be the last day of your life. Your death won't be accidental or by natural cause. You'll be murdered.

Everyday in America, our major tabloids and television newscasts are filled with frightening reports of senseless murders, rapes, robberies and batteries inflicted upon innocent, law-abiding citizens. The basic scenario, or *modus operandi,* is all too familiar: you are the prey.

* * *

Your chimes by the door jangle. The sound is friendly; it chants a sound of warmth and safety. Opening your door has become as automatic as answering your telephone when it rings, so, without another thought, you respond by running to the door and flinging it open to greet the caller.

Unfortunately, you are not a member of the *Brady Bunch,* or the *Partridge Family,* or the *Nelson Family,* or any of those saccharine-coated family shows depicting total trust and complete stupidity by whipping wide their front doors for the world to come through.

It isn't like that, as you know, but we all tend to forget—at least until something tragic happens to us or our family.

Ding-Dong, Ding-Dong.

The door looms at you as you approach it; the day's doldrums have you cocked into automatic, so you grab the knob, give it a twist and yank, sending the door open wide.

"Hi," says the younger of the two men, sort of smiling at you, hands self-consciously crammed in his pockets. "Our car won't start. Must be a dead battery. Can we call the Auto Club?"

You tell them, *sure, come right in*—you're being hospitable, helping the guys out—and it makes you feel good about yourself. Wonderful. That's exemplary; but not practical these days in these cities.

One of them walks to the telephone, the other remains by the door. You go back to your favorite chair, remote control in hand, and flick back to *Dallas*. J.R. is up to his dirty tricks again. You're relaxed, and pick up the half empty drink you left, and then—AND THEN SOMETHING BLUNT PRESSES HARD AGAINST YOUR SPINE.

"Get your ass on the ground, *now!*" Your body is slammed down onto the hardwood floor. You look up and see the other man with a gun in his hand. "Where's your money? Take the ring off your finger," the man with the blunt instrument in your back orders.

All you can think about is this can't be happening to you; it's all you can do just to swallow; you don't remove your ring, and you find yourself being kicked in the mouth. "Where the hell is your money?" you're asked again. You try to respond, but you're in shock.

The last thing you'll ever remember is a dull, high-pitched scream—your own.

Terrible, isn't it? Dying like that. You're not really dead, but you *could* actually face this situation some day. Let's turn the clock back.

Ding-Dong, goes your bell.

You don't automatically open the door. For a few dollars,

you can have the security of having a peephole in your door, whereby you'd see who was outside before you opened it!

If the person(s) are strangers, you don't open it!

If you don't have a peephole, look out the window; and if the windows in your house don't have a view of the area immediately in front of your door, you're in big trouble, because without a view of the person calling, you're gambling—that is unless you know the person—taking a chance every time you answer your door.

If you don't know the person, demand identification. But you aren't going to accept just any identification, such as a person saying he is with the gas or electric company, or telephone company, Western Union, special delivery, United Parcel Service, or any number of the well-known institutions we deal with daily. Anyone can pretend to be anyone. And some professional criminals will go to great lengths to disguise themselves to look like the phone repairperson, the gas company representative, or anyone. Even their identification can be phony.

If you want to be safe in your home, you must follow this rule: DO NOT PERMIT ENTRY INTO YOUR HOME BY ANYONE, REGARDLESS OF WHOM THEY CLAIM TO BE, UNLESS YOU KNOW THEM BY SIGHT.

If the person claims to be from a utility, call them first. Ask them if they have a Mister So-and-So, and if so, has he been scheduled to go to your house for repair work. If they say, "No," hang up and dial the police. Even if the guy is 50 years old, smiled a lot through the peephole, looked like your uncle, and wore glasses and carried a cane, ask for identification and call to verify.

It's best to be wrong, and not let someone in who perhaps is legitimately at your home to perform a needed service, than to be DEAD WRONG.

You aren't the first person to fall for the "my-car-won't-start-can-I-use-your-phone?" routine. "There-has-been-an-accident-I-need-to-call-an-ambulance" line is another way rapists and robbers get invited inside the house.

If this happens to you again, you'll know to *tell them to remain outside, and that you'll call the ambulance, the Auto Club, or the police for them.* But keep the door locked!

But suppose you let your guard down, as we all might do? What happens is you invite trouble by thinking "It's daytime, it's safe." You go to your garden and begin working on the petunias.

You've left your front door unlocked. A burglar has cased out your home and walks right through the front door. If you're lucky, he'll take only your valuables, and leave you to your gardening. But if you happen to walk in while he's stuffing loot down his jeans, he might shoot you, stab you, beat you with a heavy tool, or with his fists and feet.

Of course, the best prevention is when you're home, inside or out, keep the out-of-view doors locked.

Warning: don't think because the person at the door is a woman that any of the rules of self-preservation can be ignored. At this moment, there is a ring of criminals using females to get the person to open the front door. The others, hiding out of your view, follow her in, weapons out.

Beware of the stranger wanting to enter your home—regardless of the reason. Others have ignored this warning. It cost them their lives.

Many burglars enter residences at night. They are quiet; so quiet that there are times when victims sleep right through the burglary. Hide jewelry, money, or valuables from easy reach. The majority of burglars don't carry weapons; they know if they're caught the punishment will be increasingly severe. So, if you're awakened by a prowler or burglar, REMAIN CALM, don't put on the lights, and don't yell or scream. If you have tear gas, have it in hand, safety off and ready to use. If you can, call the police—quietly.

Again—the ordinary house burglar would have passed by your place if you had followed the safety precautions outlined in later chapters. But, if you're like most people and let a few safety measures "slide," you'll likely be the next victim.

Let the disturbed people out there raping, killing, and crippling our friends, our neighbors, and our families huff and puff and swear to blow down our lives all they want... but don't let them in by the hair of your—oh, you know the story.

Stick 'Em Up

You didn't ask to be mugged, did you? Experts claim that if you don't resist an armed robber, often you'll escape injury. This is not always true. The thief's motivation primarily is to take your money and property. Whether he beats or kills you, it's generally an after-thought. You may be murdered if the robber thinks you'll identify him. Recently in Los Angeles, a group of criminals held up a restaurant chain. The employees were put in the walk-in freezer, then the robbers decided to kill them. To prevent identification, they were shot down in "cold blood."

How would you react if someone told you, "Stick 'em up?" Unless you've been a victim of an armed robbery, you can't fully understand the trauma. Your fate could be decided by your captor. Co-author Milo Speriglio recalls facing this true-life experience:

"There were thirty customers, including myself, in a branch of Bank of America, Friday, November 25, 1960. A day I'll remember for the rest of my life. It could well have been the last day of my life. Three men in their 30's entered the bank—two carried sawed-off shotguns, the other a pistol. The fourth remained in the get-away car. They came to make a large withdrawal of $4,064, and they weren't depositors of the bank.

"One of them pushed his shotgun into my face, and ordered, 'You motherfuckers get your bodies on the floor, face down, or we'll blow you bastards away!' The tellers

were instructed to keep the palms of their hands on the counter, while the bandits cleaned out the drawers. No one panicked. "Jessie James" kept pressing his shotgun into the back of my head. I knew the silent alarm must have been triggered by then.

"For the first time in my life, I hoped the police *would not* come until the robbers left. I didn't want to be in the center of a shoot out, especially when "Jessie's" shotgun rested on my head. I also imagined being taken hostage. 'Please Mr. Policeman, have a flat tire or something,' I thought to myself. I was not the only one frightened; later I learned all of us feared for our lives, most of us saying silent prayers while on the ground. It seemed like hours, but within 45 seconds, the bank robbers were gone—before the police arrived.

We all felt relieved. The customers wanted to leave. I requested bank employees to take the names and addresses of the witnesses who didn't want to wait until the police arrived. They couldn't be forced to stay. I covered the tellers' counters to preserve any possible fingerprints. Leslie Adams, the branch manager, walked up to me and said, 'Milo, I felt much safer knowing you were in the bank.' I'm glad you felt safe, Leslie, because I didn't.

"Back in the 1960's, bank robberies were uncommon. Today, in Los Angeles alone, we have at least one a day. This robbery made the headlines of both major local papers:

*SHOTGUN GANG TERRORIZES
SCORES IN L.A. BANK RAID*

In the second newspaper, we shared the headline with another newsworthy event:

L.A. BANK BANDITS HOLD 30 AT BAY
* * *
*IT'S BABY JOHN JR. FOR
KENNEDYS AT 6 LB, 3 OZ.*

The news reports called this a 'typical wild west holdup,' and added, 'The bandit, who covered the patrons, yelled like a maniac during the robbery.' A few months later Jesse James and his Gang were arrested and convicted."

In the 1980's, all forms of robbery are wide-spread. Robbery is committed with the use of force, weapons, or both. Attempting to talk the robber out of robbing you won't work. He doesn't care if you're rich or poor, on social security, or that the doctor has told you that you have six months to live. Your best protection is not giving the criminal an opportunity to rob you. Throughout this book we give you hundreds of tips to avoid becoming a victim.

Business Robberies

Most armed robberies occur in business establishments. If you are an employee or owner, then this section should be of great concern to you. Businesses are more vulnerable to robberies because they have large sums of cash. Liquor stores, restaurants, banks and gas stations are among the most frequent targets. Cab drivers, too, are held up frequently. The least likely business to be told "stick 'em up" is a gun store. The criminals know each employee has a gun in immediate reach, and knows how to use it.

Should you have a gun in your place of business, or in your home? The answer is yes, but only if you have had adequate training to use the weapon; you know how to clean it, how and where to store it; you're satisfied that, if forced to, you can successfully use it for your protection. Later, you'll read about the pros and cons of guns. You must also know the laws about firing your weapon. You have the legal right to shoot an attacker if you, or someone else, is in *immediate* danger of great bodily harm. You can't shoot someone on your property just because he is trespassing.

Hundreds of small and large manufacturers produce an abundant number of alarms. They sell as low as $5 up to hundreds of thousands of dollars. Silent alarms are costly, and are often found in business establishments; while not affordable for many, they, too, are in residences. Any alarm is better than no alarm. Before you purchase an alarm, there are a few things you should know:

Professional alarms may deter a robbery, but a professional robber can disconnect it before committing the crime. Some of the more sophisticated alarms send a signal that it is being tampered with. The alarm will alert the police, but will they arrive in time to save your money and property, or, most important, your life? False alarms are frequent, and the police will not risk their lives, or others, to respond quickly.

Lots of businesses use the modern-age, closed circuit television to monitor areas of potential risk. These not-so-hidden cameras are regularly watched by security personnel, even in new high rise buildings. Another monitoring device is the single frame camera. It's a deterrent, but also responsible for identifying robbers. We know of a rather recent case where a liquor store clerk was murdered during a holdup, and the criminal was brought to justice after being identified on film.

Besides using technical equipment, many robberies could be avoided by following these security measures. If your business requires keeping cash and/or other valuables inside a safe, where do you think the safe should be kept? In the back of your establishment, hidden from the view of the potential robber, or near the front window? You're wrong if you picked the first option. The safest place for your safe is somewhere exposed to public view—from both inside and out. This is assuming the safe is a large one.

Large plate glass windows offer you protection from robbers. If you have them, don't keep the windows draped and out of view from passerbys. Have you ever walked or driven by a bank, and couldn't see inside?

How well do you protect your cash? If your daily deposits are quite large, don't risk taking them to the bank; use an armored car service. Checks written to your business can be cashed by the robber. Instruct your employees to stamp the "for deposit only" endorsement as soon as received. Today, the "plastic card" is used nearly as often as cash or checks. The credit card voucher is worthless to the robber, but if he takes them, you don't get reimbursed from the bank. Treat them like you do your cash.

One final advice to businesses. Nearly every business firm makes this mistake everyday. The 50's, 100's, checks, traveler's checks, and credit card vouchers are placed underneath the main cash drawer of the cash register. You do this to "hide" the big money, but every holdup artist knows this "trick." When you're ordered to "Stick 'em up," the robber will clean out the entire cash register. Top and bottom. To prevent this from happening to you, frequently remove the big bills, checks, and credit card vouchers from the cash register, and really hide them, until it's time to make the deposit.

Street Robbery and Muggings

At the top of your lungs, you scream for help, but in vain. You can't believe this has happened to you; your body is trembling as you pick yourself off the ground, and you're in great pain. But you wipe the dripping blood from your mouth, nose, eyes and ears, and you're relieved because the muggers are gone; the punishment is over.

The muggers really enjoyed punching and kicking you. They had their "fun" and they also have your money and jewelry. In the emergency room at the hospital the doctor tells you, "Your broken ribs will heal within a few weeks, the contusions on your head are minor, there's no fracture of the skull, and those bumps and bruises will soon disappear." But there will always be a mental "scar" to "mark" this occasion.

You didn't ask to be mugged—or did you? Larry is nearly seventeen, and he's on drugs, and to support his habit, he robs, and sometimes mugs. He was arrested for his last robbery. He tells us about that night: "I pick and choose who my score (victim) will be, I wait and watch, I look for two things: the right score, and the right time. I stand hidden behind some bushes with my gun." Larry described some of the potential victims that passed him by:

"I seen this guy about my age, long hair, beard and moustache, wearing jeans, and then this teenager went by carrying a bunch of school books, and several guys were jogging together, then this man, in an expensive tailor-made suit—and he had his shirt open—and his gold chain could have choked a horse ..."

Guess who Larry robbed? As it turned out, the man with the gold chain was an off-duty police officer. Larry went to jail; a "guest" at the juvenile detention center he calls summer camp.

Larry's newly acquired friends listen to him as he boasts about his life of crime. "You know, they call it the city, but it's really a jungle. It's survival, guy. We are the animals—you like to call us animals—and you, you're the prey. We're hungry. We need your money to buy our drugs. We'll stalk you, and when the time is right, pounce on you stupid bastards, and take what we want, and kick the shit out of you. I remember this one time, this dumb broad wore a ring so tight, I couldn't pull it off, so I had to cut her finger off. But I got the ring—sold it for forty bucks. It's a good thing for her she didn't have no more rings on!"

Sometimes you're mugged and not robbed. Some juveniles get their "kicks" out of beating people up. At this very moment, in Los Angeles, organized gangs are responsible for more homicides than any time in the city's history (Los Angeles just had its bicentennial). Avoid, at all costs, being in the area where juveniles, particularly gangs, congregate.

Larry made us aware that the city is a jungle. The muggers are the hunters, and we are the prey. Law-abiding persons are *open season* to the muggers and robbers. The creatures of the jungle do most of their hunting at night. Muggers also operate in broad daylight, and they often work in groups, just like a pack of wolves. Like with the con men, which we later expose, one of the "pack" diverts your attention, generally by conversation, while the rest of the wolves just wait, ready to attack.

You're asked:

"What time is it?" ... "Have a Match?" ... "Where is South Street? ..." If you look down at your watch, and give them the time, that watch may soon be theirs. Give them the light, or directions, do anything they ask, and within seconds the wolves from the jungle will pounce on you.

How do you know if the person talking to you really is a criminal? You can't go around for the rest of your life suspecting everyone is a "wolf." If you're about to be mugged or robbed, there's a good chance you'll know, just by the person's tone of voice, his conversation, facial expression, eyes, gestures, and body movements. Often, an inexperienced mugger or robber will act nervous.

In this jungle world we live in, we must be aware that our only protection is self-protection; we must do whatever we can to prevent the crime, and defend ourselves from being attacked. Take the tiger, the wolves, and the lions by the tail, and let them know we're no longer easy prey. Within the pages of this book, you'll find ways and means to protect yourself.

Residential Robberies

If you are going to be robbed, the least likely place will be in your home. Less than twenty percent of all robberies occur in the residence. You still should be very concerned. Co-author Thomas Eubanks recalls a rather recent armed robbery of a friend, and it almost cost him his life. It didn't start in the house, but the crime was committed there.

"My friend won $96,000 in *cash* at a pyramid game, recently very popular, but illegal. The way it goes is this: you put up $1,000 and attract friends and family, and soon the pyramid grows. The one in first place takes home the big prize. Most persons lose, a few win. Frank was a winner and went home with almost 100 thousand in cash in his pocket. He was so happy he cried. After this big score, he drove home. He didn't know it, but he was followed.

"As soon as he put his key in the front door, he felt the barrel of a gun in his back. The robbers forced him inside, and removed his winnings. He was lucky they didn't kill him."

Following you home, robbing you as soon as you enter your door, is common for armed robbers. Inside your home, you should be safe. But you aren't, because the robber tricks you into letting him inside. His "tricks" are reported elsewhere in this book.

With your doors locked, windows closed, you feel secure. But we all go outside everyday. The robber knows our habits. This is one of the latest methods he uses: you wake up, throw on a robe, and, planning to read the news during breakfast, go outdoors to pick up the morning newspaper; as you bend down to pick up the paper, Mr. Robber greets you, gun in hand, and becomes the bearer of "today's bad news." He escorts you back into your secured house. While he robs you, you probably want to cry "the bastard caught me with my pants off!"

Rape: When, Where, What and Why

Except for being murdered, rape is the most serious violation of your body. It is the fastest growing crime in America today. Men as well as women are victims of rape.

The day could come when you're saying "I can't believe it happened to me." Each year thousands of you fall prey to the rapist. What follows is a profile on rape based upon national crime statistics of reported forcible rapes.

As a potential rape victim, you can feel a little safer during January and December of each year because less rapes occur during these months than the other months. But be extra alert during August—the rape month of the year.

During the hours of 8:00 P.M. and 2:00 A.M., 70% of all rapes occur. You are least likely to be raped or attacked between eight in the morning and two in the afternoon. However, regardless of the time of day or the month of the year, a forcible rape is reported every seven minutes of everyday.

More than half of these crimes occur beginning on Friday evening and on through Sunday.

Where do most forcible rapes take place? You would think you'd be safer at home than in the streets. The fact is, there is a better chance of being raped in your own residence. Approximately one-fourth of all rapes are committed in the streets away from home. The third likely place is in a motor vehicle—probably your own.

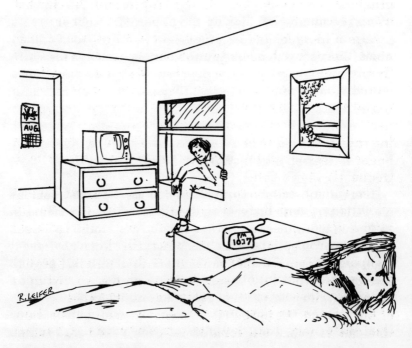

Some rapists have been ten years old. Some even younger. Senior citizens commit rapes. But the average age of an attacker is 16 to 20 years old.

The majority of all victims never personally knew their attacker prior to the rape. One-third of them had some casual acquaintance with the rapist. And an estimated 10% were friends of some sort with the rapist, while nearly 7% of the forcible rapes were committed by relatives. Nationwide, nearly 12% of all women raped were hitchhiking at the time.

What would you do if someone was about to rape you?

Well, about one-third gave no resistance. Some 10% attempted to run. And about 20% made some type of physical resistance. In approximately half of all rapes, the common resistance was only verbal.

In 20% of the reported crimes, both the victim and the offender were under the influence of either alcohol or drugs. In nearly half of the reported rapes, only the attacker was under the influence.

Does a man, by verbal threat and physical appearance alone, demand and receive and abuse your body at his will? Does he just rape you as he pleases? Based upon national statistics, yes, he did, in nearly fifty percent of all rapes. In the other half, the rapist used force by physical strength or used a weapon. The most common weapon employed by the rapist was a gun or a knife. In cities with a population under 100,000, the knife or sharp instrument was used more frequently than a gun.

Every victim of rape receives some kind of injury. Almost all will carry with them an everlasting mental scar. Nearly 25% suffer injuries severe enough to require medical attention or hospitalization. Some are killed.

Often, the rapist demands far more than only the sexual intercourse. One out of every four will force the victim to perform oral copulation. To further abuse the victim, in 7% of the attacks, the perpetrator makes an anal penetration. One out of every ten forcible rapes consists of vaginal intercourse, oral copulation and anal penetration. Other forms of sexual acts are also included in the attacks on one out of every twenty rapes. Frequently, these acts are repeated over a period of hours, sometimes days.

Some people still believe that if a female dresses in permissive, suggestive clothing, she is just asking for it. This is an ignorant viewpoint. If a woman carries a purse, is she "just asking" for someone to snatch it?

Physical attractiveness and age have no bearing upon the sick rapist. He has been known to physically rape infants as well as persons in their 90s.

Odds are about a million to one that the rapist will be a doctor, a dentist, a lawyer, or a Certified Public Accountant. Only 10% are professionals and white-collar workers. Forty-five percent of these crimes are committed by so-called blue-collar workers, 30% by the unemployed, and 15% by students.

Rape is one of the most repeated crimes of all. Few criminals are caught committing their first rape. Many rapists have committed a score of forcible rapes. If you become a victim, report it. Try to describe your attacker. You might prevent another from becoming a statistic.

From our "rape profile" we have determined that 23% of the rapists arrested had had previous rape arrests. Among them, 25% had records for other sex offenses, and more than one-third had records for other violent crimes.

Rape is a four letter word. And your only protection is self-protection. Do whatever you can to prevent being violated. If you live in California, or any other state which permits it, you can be licensed to carry a tear gas weapon for self-defense.

Unless you have been a victim of forcible rape, you can't fully understand the trauma. In one-fourth of all rapes, in fact, there are two rapists violating you. In 12% of the reported rapes, there are three or more.

The Federal Bureau of Investigation reports that one out of every three females in the nation will be assaulted or raped in her lifetime.

Are you prepared?

VIOLENT CRIMES

Approach Your Car With Caution

Scott watches you head for the market. You leave your avacado green Buick Regal parked behind some trash bins offering sufficient cover.

He scuttles through some parked cars, and, upon reaching the Buick, unpacks a red-handled ice pick from his coat pocket. Placing the point about an inch above the keyhole in the door, he firmly pounds the handle with a palm. The ice pick pierces the door; with a skilled, blind flick, Scott pops up the lock rod.

After a brief scan of the parking lot for onlookers, Scott jumps into the back seat, and almost simultaneously flicks the domelight switch to "off." A quick pry on the plastic cover with his ice pick, and off it pops. He removes the light bulb.

In the safety of darkness, he cradles himself on the floor, and waits.

* * *

You hate to shop. But your daughter who usually does your shopping for you is skiing in Lake Tahoe, so you have only twenty minutes to get a few snacks before your good friends, Heather and Michelle, arrive at your condominium for a night of Three-handed Spades.

You grab a few items and rush out of the market, carrying your bag of food. Your mind races: "Did I get everything? Is there enough cream cheese at home, or should I go back and get some more?"

You spot your car among the horde of others. "I hope Michelle brings that extra deck of cards." You approach your car, slipping in the key, turning, thinking to yourself, "If she doesn't, I can always come back and get more cards, but now that I'm here, I might as well—no, she'll bring them. I don't have the time anyway." You push your door open with your knee, struggling with the grocery bag, and note right off that the light is off. "Fuse is out. Have to fix it tomorrow."

Setting the grocery bag on the front seat, you jump in behind the wheel and slam the door, checking your watch. "Only ten more minutes, dammit," you think to yourself.

The engine revs. Into gear and out of the parking lot you go.

You suddenly feel the horrible cold feeling of something hard and sharp against your throat. "Don't turn. Keep drivin'. Or I'll *push* this thing right *through you.*"

* * *

If you're the average person, you'll enter your car, after parking at the market, at work, when shopping, visiting, and other times, an estimated 650 times per year, 27,795 in a lifetime. You are probably unaware of the risk you take each time you approach and enter your car.

The bad guy knows the best time to get you is when you're alone, approaching your car, especially in isolated, dark areas; but often in bright daylight, the car still remains a favorite method for taking a victim.

The two ways criminals use cars are:

1. While your car is unoccupied, they enter it. Regardless of how good you think your car's security is, even if you have a sophisticated alarm, for the expert, it's no problem to break in within 15 seconds.

He may tape the door button or extract the lightbulb from the domelight—either way, when you open your car door, everything is dark.

Once you get in the car, you're his. If you're a man, he may rob you; if you're a woman, rape may also be included in the attack against you. After raping you, he may take you somewhere secluded and kill you. It happens.

2. Until recently, the *modus operandi* described in method number one was the most popular. Today, however, Bad Guy has a more effective approach; often his "M.O." is not to enter your car. What he does is this: he slips underneath your car, hidden in the darkness. As you approach and begin

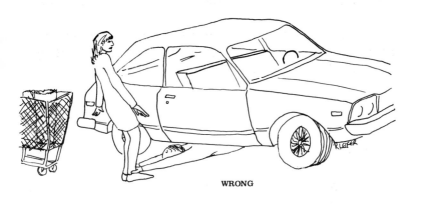

WRONG

RIGHT

to unlock your car door, he suddenly grabs your ankles and yanks your legs, causing you to fall, where he can do with you as he pleases.

Some are sicker than most. Rapists in the last few years have also added one other affliction upon the victim. Instead of yanking on your ankles from under the car, he'll skillfully slice your Achilles tendons with a razor blade as you open your door. With the backs of your ankles cut, you'll fall, in great pain, and the rapist, at that point, has you under his power.

Let's turn back the clock a few minutes. Could this terrible crime have been prevented? Could you have done something before getting back into your car? The answer is *yes*.

Approach your car with caution. Before you get too close, though, shine a pocket-size flashlight under it, and if Bad Guy isn't under it, before you unlock your door to get in, shine your flashlight into the front and back seat, or even into the campershell of your camper, before climbing in behind the steering wheel. If you spot somebody in your car or under your car, run and scream, "FIRE!" not "rape" or "robbery"—just "FIRE!"

Furthermore, if he chases you, and you have tear gas, the chances are you'd be able to temporarily disable him. If he catches up with you, and you haven't bothered to get tear gas yet, use anything to defend yourself. (Self-defense techniques will be explained in Chapter Four.)

One final warning: Bad Guy doesn't always hide in or under your car. He might be behind another car parked beside yours; he could hide in a bush, in trash bins, even under a car adjacent to your own.

So be ready. Your tear gas, if you have it, should be in your hand, ready to fire. The first big step along the road to self-preservation, must be SELF-PROTECTION, and self-protection can be accomplished through AWARENESS; so be conscious of yourself, and not just whether or not you have enough cream cheese.

Myths, Misunderstandings, And Categories Of Rape

Awareness also means that you be conscious of possible threat. But what does a rapist look like? And why would he want to rape you? Men don't rape women because they act or dress seductively. Most rapists don't seek sexual gratification from the act.

Admitted rapists often prefer attacking the victim because they enjoy overpowering her. They are not sex-deprived. Most have consenting, available relationships with other females.

Rape, most often, is not a sex crime but an aggravated assault. One of the most dangerous rapists is the repeater. He'll rape a score of females before he is arrested. After spending time in prison, he's free and back on the streets. He rapes again and again. This type of person is more likely to kill his victim to prevent identification.

Another dangerous type is the rapist who wants to dominate and degrade his victim. Often, he uses physical force, and in many cases, he can't have an erection, or finds it impossible to ejaculate.

There are times he rapes for sex. This type usually has no criminal record, nor did he intend to rape. You meet this type of person many times at social functions. And leave with him. He is often a married man out for a good time. He forces himself on you even though you say "no." He may have had too much to drink, or he's "over-sexed," but whatever the reason, you are his, and if you don't consent, he'll take you. He commits his first rape. Most likely, you'll live to tell about it.

The next classification of rapist is also a "non-criminal" type: he is your stepfather, natural father, uncle, brother, relative, former spouse, someone you hardly know, the grocery delivery boy, the neighbor, or whomever. The motivation in this case is strictly physical attraction.

The most dangerous of all rapists is the strangler type.

Most prominent of all was the Boston Strangler. This type of person is often incapable of caring about the sex act, he just wants to rape and torture. He may use a broom or other "instruments" to penetrate the vagina. He sexually abuses his victim.

Regardless of the sex performance, the victim often receives the ultimate conclusion. Death.

Knowing now what you're up against, do you submit or resist?

Submit Or Resist?

Two recent studies founded by the National Institute of Mental Health revealed that the more a woman resists a rapist, the more likely she'll avoid being raped.

For years, many "experts" advocated submission to the attacker to prevent injury. It has now been proven that the females who are passive or cried, or attempted to talk the rapist out of performing the act, or begged for sympathy, or just plain submitted, were found to be more likely the candidate for a sexual assault.

One noted self-defense expert, Mary Conroy, who teaches a class on the technique in resisting rape at California State University in Los Angeles, was quoted as saying, "Rapists are not amenable to reason. For the most part, a woman who resists does not increase her risk of serious injury. She may suffer from a few bruises, but much less than a woman who is compliant."

The word is this: "Kick them between the legs. Gouge out the rapist's eyes with your keys clutched in your hands. Slash him in his Adams' apple. Scream. Use any means of self-protection." If you have a tear gas weapon, this may be your best defense.

But unless you're overpowered, defenseless or in a situation where escape is impossible, don't submit.

Women who live in or around the following forty cities should take extra precautions. Men, too.

VIOLENT CRIMES

The 40 Most Dangerous Cities For Females

Men are often raped by other men. Not just in prisons, but in the streets, their homes and motor vehicles. These rapes are not included in the following statistical information. Such rapes are seldom reported to the police.

Many attempted rapes of a female go unreported. Several years ago, it was estimated that only 1 in 5 forcible rapes were actually reported; today, with the new laws that protect the victims, around half are reported.

Based upon our independent research, you can multiply the number of reported rapes below by six to determine the actual number of attempted rapes, and double the figures to establish how many actual forcible rapes occurred.

If you live in, or nearby, any of the following cities, you're a likely potential victim of rape. This list of cities, followed by the number of reported forcible rapes, is what we call the Nation's Top 40.

LATEST AVAILABLE *ANNUAL* STATISTICS
OF REPORTED FORCIBLE RAPES

1. New York, New York 3,875
2. Los Angeles, California 2,508
3. Chicago, Illinois...................... 1,655
4. Houston, Texas....................... 1,481
5. Detroit, Michigan..................... 1,369
6. Dallas, Texas........................... 983
7. Philadelphia, Pennsylvania 838
8. Memphis, Tennessee 704
9. San Francisco, California 664
10. Atlanta, Georgia 656
11. Denver, Colorado....................... 626
12. Cleveland, Ohio 612
13. Baltimore, Maryland 564
14. St. Louis, Missouri 555

15. District of Columbia
    Washington, D.C. 489
16. Phoenix, Arizona 477
17. Boston, Massachusetts 464
18. Indianapolis, Indiana.................... 439
19. Kansas City, Missouri................... 436
20. Portland, Oregon 435
21. Columbus, Ohio 423
    New Orleans, Louisiana 423
22. Seattle, Washington 421
23. Newark, New Jersey..................... 417
24. San Jose, California 407
25. Oakland, California 373
26. Jacksonville, Florida 372
27. Fort Worth, Texas 345
28. San Diego, California 331
29. Minneapolis, Minnesota 327
30. Oklahoma City, Oklahoma 311
31. Tampa, Florida 308
32. Birmingham, Alabama 285
33. Milwaukee, Wisconsin................... 283
34. Cincinnati, Ohio........................ 282
35. Long Beach, California 278
36. Nashville, Tennessee 266
37. Buffalo, New York...................... 265
38. Pittsburgh, Pennsylvania 263
39. Miami, Florida 261
40. Las Vegas, Nevada 254

Other cities that had 200 or more reported forcible rapes include, Toledo, Ohio; Sacramento, California; Honolulu, Hawaii; Gary, Indiana; and Charleston, West Virginia.

We researched the statistics of every city and town in the United States with a population of 10,000 persons or more. Our investigation established that only 448 cities, out of 2,497, had no reported rapes during the last national crime report.

VIOLENT CRIMES

If you *are* raped, here's where to call for help . . . .

Rape Crisis Hotline Numbers

The following is a list of telephone numbers throughout the nation where rapes occur most frequently. The numbers may change. Many are answered twenty-four hours a day by professionals.

Some major cities have no rape crisis centers. If the area you live in isn't listed below, call telephone information. The name of the service may vary:

        Rape Crisis Centers
        Women Against Rape
        Rape Relief Center
        Rape Counsel Center
        Rape Call
        Rape and Assault Center
        Rape Awareness Center
        Sex Abuse Center
        Rape Hotline
        Crisis Intervention Bureau
        Rape Intervention

If you can't locate a rape crisis center, call a community hospital or your local police for the center nearest you. The women at the rape crisis centers are volunteers. Many of them have been victims themselves. They can give you comfort and aid, so if you are a victim, call them:

ALABAMA
 Birmingham (205) 323-7273 (Rape Response)
ALASKA
 Anchorage (907) 276-7273
ARIZONA
 Phoenix (602) 257-8095
ARKANSAS
 Little Rock (501) 375-5181
CALIFORNIA
 Downey (213) 868-3783
 East Los Angeles (213) 262-0944
 Laguna Beach (714) 494-0761
 Long Beach (213) 597-2002
 Los Angeles (213) 793-3385
 Sacramento (916) 447-7273
 San Bernardino (714) 883-8689
 Santa Monica (213) 392-8381
 Ventura County (805) 529-2255
 Riverside (714) 686-7273
COLORADO
 Boulder (303) 443-7300
CONNECTICUT
 Hartford (203) 522-6666 (Sexual Assault Center)
DELAWARE
 Wilmington (302) 658-5011
FLORIDA
 Daytona Beach (904) 255-1931
 Jacksonville (904) 354-3114
 Miami (305) 547-7810
GEORGIA
 Atlanta (404) 659-7273
HAWAII
 Honolulu (808) 947-8511
IDAHO
 Boise (208) 345-7273

ILLINOIS
   Chicago (312) 372-6600 (Chicago Against Rape)
      (312) 883-5688 (Rape Victim Advocate)
      (312) 744-8418 (Rape Victim Emergency)
   Peoria (309) 674-4443
INDIANA
   Indianapolis (317) 630-7791
   (Community Mental Health)
      (317) 353-5947 (Crisis Intervention)
IOWA
   Des Moines (515) 286-3838
KANSAS
   Kansas City (316) 923-1123
   Wichita (316) 263-3002
KENTUCKY
   Louisville (502) 581-7273 (Rape Relief)
LOUISIANA
   Baton Rouge (504) 383-7273
   New Orleans (504) 821-6000
MAINE
   Portland (207) 774-3613
MARYLAND
   Baltimore (301) 366-7273
MASSACHUSETTS
   Cambridge (617) 492-7273
MICHIGAN
   Detroit (313) 872-7273
   Flint (313) 767-2662
MINNESOTA
   Minneapolis (612) 825-4357
   St. Paul (612) 777-1117
MISSOURI
   Kansas City (816) 923-1123
   St. Louis (314) 725-2010
MONTANA
   Billings (406) 259-6506

NEBRASKA
  Lincoln (402) 475-7273
NEVADA
  Las Vegas (702) 735-1111
  Reno, Sparks, All of Nevada (toll free)
  1-800-992-5757
NEW HAMPSHIRE
  Manchester (603) 668-2299
NEW JERSEY
  Cherry Hill (609) 667-3000
              (609) 858-7800
NEW MEXICO
  (505) 982-4667
NEW YORK
  Manhattan (212) 777-4000
NORTH CAROLINA
  Charlotte (704) 372-4357
NORTH DAKOTA
  Fargo (701) 235-7355
         (701) 293-7273
OHIO
  Akron (216) 434-7273
  Cincinnati (513) 381-5610
  Cleveland (216) 391-3912
  Columbus (614) 221-4447
OKLAHOMA
  Oklahoma City (405) 524-7273
  Tulsa (918) 744-7273
OREGON
  Portland (503) 235-5333
PENNSYLVANIA
  Harrisburg (717) 238-7273
  Philadelphia (215) 922-3434
  Pittsburgh (412) 765-2731
RHODE ISLAND
  Providence (401) 861-4040

SOUTH CAROLINA
  Greenville (803) 232-8633
           (803) 271-0220
SOUTH DAKOTA
  Sioux Falls (605) 334-6645
TENNESSEE
  Memphis (901) 528-2161
  Nashville (615) 327-1110
TEXAS
  Dallas (214) 521-1020
  Houston (713) 228-1505
  San Antonio (512) 231-2636
UTAH
  Salt Lake City (801) 532-7273
VERMONT
  Burlington (802) 863-1236
VIRGINIA
  Norfolk (804) 622-4300 (Rape Information)
WASHINGTON (State of)
  Seattle (206) 632-7273
WEST VIRGINIA
  Charleston (304) 344-9839
             (304) 344-9834
WISCONSIN
  Waukesha (414) 278-4617
WYOMING
  Cheyenne (307) 632-2666
WASHINGTON D.C.
  (202) 232-0202

You students should keep this in mind . . . .

How Safe Are Universities and Colleges?

Thousands of hours of research were needed to compile the data for this report. Without the assistance of the Federal Bureau of Investigation, this information could not be assembled.

We know that we are not safe at home, in the streets or at work. But how safe are we at a university or college, protected by campus police, surrounded by faculty and fellow students? What follows are current available statistics of actual crime reports made annually. Many crimes on campus are never reported, particularly attempted rapes, assaults or forcible rapes.

LAST YEAR'S REPORTED VIOLENT CRIMES

University	Count
Rutgers University, New Jersey	76
University of Illinois, Urbana	50
University of Florida	48
University of Maryland, College Park	46
University of California, Berkeley	45
University of Oklahoma	44
Kent State University, Ohio	39
Massachusetts Institute of Technology	37
Michigan State University	37
Northern Arizona University	36
University of Houston, Texas	32
Boston University, Massachusetts	31
Ohio State University	31
University of California, Santa Barbara	30
California State University, San Jose	29
South Illinois University, Carbondale	27
South Illinois University, Edwardsville	25
Townsend State University, Maryland	25
Northeastern University, Massachusetts	25
Colorado State University	24
Illinois University, Bloomington	24
Montclair State College, New Jersey	24
University of Arkansas	23
Florida A&M University	23

VIOLENT CRIMES

University of Massachusetts, Amherst 23
Cheyney State College, Pennsylvania 22
Boston College, Massachusetts 21
University of Alabama 20

REPORTED FORCIBLE RAPES
Indiana University, Bloomington............. 11
Southern Illinois University, Carbondale.......8
Rutgers University, New Jersey 8
Ohio State University 8
California State University, San Jose...........6
Townsend State University, Maryland..........6
University of Massachusetts, Amherst 6
Michigan State University.....................6
Northwestern Oklahoma State University......6
University of Southern Florida 5

The rest of the universities and colleges had four or less forcible rapes. Many of these crimes occurred off-campus, and are not included in the above figures. Based upon our independent interviews with students throughout the nation, the number of reported rapes is equivalent to only a small fraction of actual attacks. Many campuses provide escort services to female students when walking off-campus.

MURDER OR NON-NEGLIGENT MANSLAUGHTERS

This crime on campus is always reported. Each of the following had one or more murders or manslaughters during the period reported in the last available F.B.I. Uniform Crime Report.

California State College, Dominquez Hills
California State Polytechnic University,
   San Luis Obispo
Indiana University, Indianapolis
Michigan State University
University of South Carolina

HIGHEST CRIME RATE ON UNIVERSITIES AND COLLEGES

Most Aggravated Assaults:
 University of Illinois, Urbana 4
Most Robberies:
 Rutgers University, New Jersey 29
Most Property Crimes:
 Wright State University, Ohio 2,066
Most Burglaries:
 University of Maryland, College Park 551
Most Larceny and Thefts:
 Ohio State University 1,719
Most Motor Vehicles Stolen:
 Massachusetts Institute of Technology 131

Spotlight For The Bad Guy

Where do you think you would be safer? Walking down a dark, lonely street, or under the safety of a well lit area? If you picked the latter of the two choices, you missed. Recent studies are inconclusive as to how much street lighting really deters criminal activity. While well lit streets may make the residents and pedestrians actually feel safer, statistics show that crime increased in many well lit areas. This statement of fact was not true a short time ago, but perhaps there is a new trend in the criminal mind for the 1980's.

Does this "spotlight" attract the criminals like moths? Probably not. The logical answer may be that the bad guys, like everyone else, see better in the light than in the dark. Since the job of breaking into places is now easier, why not shed a little light on the subject? One thing is certain. The muggers and burglars who prefer the cloak of darkness may soon retire. And if they don't retire on their own, perhaps your local District Attorney will retire them for you.

VIOLENT CRIMES

Mr. D.A.

If you're a victim of a crime, you may come in contact with the District Attorney. He's the Attorney for the People.

The "D.A." and his Deputies initiate investigation into and prosecute all felony crimes and some misdemeanors. The City Attorney handles the other criminal matters.

The District Attorney's primary function is to prosecute those accused of a crime, and to do so in the interest of delivering justice. Many accused criminals aren't tried in court because the D.A. rejects cases that lack sufficient evidence to prosecute the suspect. Unless he's assured of a solid case, a chance for conviction, many guilty persons and accused innocent persons, too, are let go and don't face trial.

In Los Angeles County, home of the nation's largest prosecutor's office, there are 520 Deputy District Attorneys responsible for serving a population of almost eight million. Good, hard-working people.

We just hope you don't have to oppose them officially. The District Attorney also prosecutes those who commit crimes against police officers.

Criminals Kill Police Officers Too

If the police are not safe on the streets, how can *we* expect to be? We compiled the following from the latest available FBI Uniform Crime Reports, and other independent research. We discovered that 1,143 law enforcement officers in the United States lost their lives in the line of duty within a nine year period.

Today, many city and state budgets are prohibiting police officers from working as a two officer team in a patrol car. This is most prevalent in California. We found that 42% of all officers slain during the period in question were alone or unassisted. If they had a partner, we estimate perhaps one-third of these officers would be alive today.

How were the officers killed?

In 94% of the cases, firearms. Most deaths were caused by the use of a handgun. A rifle or shotgun was used in about 17% of the crimes. A knife, bomb or vehicle caused less than 5% of the assassinations on police officers.

Who kills law enforcement officers?

We researched the deaths of officers during the past decade and found 97% of the time they were killed by a male: 51% white; 47% black. The ages of the assailants ranged from 15 to 86, with the average age at 29. From those arrested and convicted, 43% of them had been arrested in the past for suspicion of committing violent crimes, and 16% of the police-killers were either on parole or probation at the time they killed the officer.

* * *

In the motion picture *The Godfather,* the killing of a police officer was the worst crime a person could commit. But today, the man behind the badge and uniform has little or no respect from the criminal: killing "the pig" has become as common as butchering hogs. Police officers know they have but one friend, a fellow officer, who'll back them up, and pursue any assailant.

Kill "a pig" and the criminal can expect to be caught; it may take just days, or weeks, or months, or even years, but the odds are they'll be brought to justice. Commit a burglary and the odds are one out of 950 the criminal will ever be arrested; kill a police officer, and the odds are better than nine out of ten he'll be caught.

Your Child May Be Molested

"Mommy, I was just outside skating, having a lot of fun, when this nice man stopped his car and asked me where a gas station was. I told him it was just down the street, but he was lost, and asked me to show him how to get there.

"Mom, I know you told me never to go anywhere with a stranger, but he was nice and he was lost. But then he *wasn't* nice; he made me pull down my pants ... oh, God, mommy, it still hurts."

This could have been your child. Cindy is only seven years old. She's just one of the hundreds of thousands of children molested each year. This type of crime seldom is reported in the media. Only one out of every ten molestations are reported to the police. The offender is almost always a male; frequently he's married and has children of his own.

Who molests children? Sick criminals who call it "child love." They're generally referred to as pedophiles. It happens everyday, and in every neighborhood, including yours.

James S. is a child molester. He was arrested and sent to a state mental hospital for cure, not punishment.

He told us, "It's easier to make love (molest) with boys, rather than girls. They (boys) are easy prey. There is a better chance to lure them away from their families. Take them fishing, swimming, to the park, whatever. The boys are looking for someone as a father image. I talk to them on their own level. I give them something they really want. I give them sex." For James, like his sick counterparts, sodomy and oral copulation with children is their way of life. For the female victims, the vagina is often penetrated.

A group calling themselves the Rene Guyon Society advocates legalizing sex with children. This "group" claims a membership of five thousand. Worst of all, this is only one of many such societies. The slogan for one of them is, "Eight ain't too late."

Sex with children is also a multi-million dollar business today. Thousands of photographs depict children having sex with other children, as well as adults. This form of "art" is found in kiddy porn magazines. It's also captured in motion—8mm film and video tapes. Child molesters told us they like to take photos of the children they have sex with, so if the child says they're going to *tell,* the molester threatens to show the photos to the victim's friends. Or they tell the child that their mommy will spank them if she ever sees the pictures.

Shirley W. is the mother of a molested child. "I warned my girl over and over again, 'do not talk to strangers, do not go with strangers.' I wouldn't even permit her to open the door to a stranger. Children are so vulnerable. They don't really understand what can happen until it actually happens to them." She told us that parents must be very serious when they warn children. She added, "Tell the child, 'The bad man can kill you, and you'll *never* come home again'."

In ten percent of the known child molesting cases, the victim was badly injured, or killed. Only one percent of the molesters are ever arrested, and this usually happens after they've committed multiple crimes. James, the child molester we interviewed, boasts of having over 200 sex acts with children before the arm of the law caught up with him. Of those arrested, many are never convicted. Why? Because often the child is too young to testify, or a good attorney will confuse the victim. One molester said, "I don't care if it's a boy or a girl, just as long as they're immature and they fit my needs."

Child molesters find their victims anywhere and everywhere; at the beach, in the streets, at schools. Some "sweet talk" the child into going with them, others have the parents permission to be with the child. Some parents even prostitute their own children. Studies show that 80% of the child molesters were molested themselves as a child, often

VIOLENT CRIMES

by a relative.

S.C.A.M. is an organization to stop child molesters. Their acronym means Concerned Citizens for Stronger Legislation Against Child Molesting. Their objectives are the same as most parents, but their unity is small in number; the Rene Guyon Society, which openly advocates legalizing sex with children, has a larger membership than S.C.A.M.

We must demand that our state and federal legislators pass very strong laws dealing with the sex offender. In the solution chapter at the end of this book, we outline extremely stiff penalties for child molesters.

Many convicted child molesters only receive probation, and they're back on the streets again raping our children. Sam P. is another child molester, and he said, "I have tuberculosis, and there's no cure for this disease—all they can do is *arrest* it. And all society can do to child molesters is *arrest* them; keep them behind bars. In prison there ain't no children."

If you're a parent, grandparent, aunt or uncle, you must realize that child molesting *does* exist; it's a crime often repeated on the same victim a hundred or more times. Be extremely cautious of who you leave your child with; that nice old man, who offers to babysit, may have other plans on his mind. Sometimes relatives are the molester. In the rock musical, *Tommy,* we saw Ringo Starr as Tommy's "wicked Uncle Ernie," who wanted to "fiddle about, fiddle about," whenever Tommy came to stay with him.

What can we do? While your child plays outdoors, give him or her a loud whistle tied around their neck. Instruct them to blow the whistle when someone tries to lure them away from their home. If your youngster is around the age of ten or more, purchase a "Shriek Alarm." Let them carry it when bike riding, while in the movies, etc. While at the movie theatre, youngsters should know to go in groups to the snack bar or toilet.

As a parent there's not much more you can do to protect

your children. Awareness is extremely important, and your children must be frequently *reminded* of the danger they face.

Perhaps the best documentary ever filmed about the sexually exploited child is *The Youngest Victim*. It aired as an ABC-TV special, September 7, 1981. Robert Blake hosted the special, and summed it up saying, "We can't isolate our children completely, we can't make them absolutely safe, but with our love and affection, they won't be such easy marks for the molester."

Compensation For Victims Of A Violent Crime

As you approach your home in the darkness, two men appear from behind your bushes.

"Don't make a sound," one orders; you obey when you spot a gun in his hand.

You're forced inside your home. They remove your jewelry; you resist and a fist smashes your mouth and jaw, joggling a few teeth loose.

They take the jewelry, but demand more. They want your money and other valuables now, and you comply.

They leave. And you're fortunate to still be alive, although minus your personal treasures. The ambulance takes you to the emergency room of the local hospital, and doctors find your jaw broken.

The hospital bill and medical follow-up cost several thousand dollars; you also lose a few weeks of work.

If your assailants are ever caught, and that is unlikely, and even if they're arrested, you can file a civil suit for damages. It may take as much as five years waiting for a trial, but if you're awarded a judgment, and *if* the criminals are still around, you then have to hope they have sufficient assets to pay the judgment.

Many victims of violent crimes are never compensated even for their out-of-pocket cost. In 1965, California became the first state in the nation to compensate these victims. Later, 27 other states followed with similar programs. This isn't an insurance policy; no one pays a premium. It's a program designed to provide temporary monetary assistance to those victims of a *violent* crime, such as rape, robbery, aggravated assault, and murder.

If you take your assailant to civil court you might collect for pain and suffering (personal injury) in addition to medical and loss of wages. The State Victims Compensation doesn't pay for pain and suffering.

In California, the average award given to victims is $2,200. The maximum their survivors can receive is $23,500 from the State. Medical bills up to $10,000 and funeral expenses up to $2,275 may be paid. If you were off work and lost wages or support resulting from a violent crime, you can be paid up to $10,000.

When your injury prohibits you from continuing your normal occupation, the State can rehabilitate you, and give you training in another field in which you can function. In California, up to $3,000 will be provided. Another $500 may be granted for legal expenses.

To find out if your state compensates its victims of violent crimes, ask your local law enforcement agency.

The front runner of this state assistance, California, is running into financial trouble. In April, 1981, the $6.4 million program found itself $1.7 million under budget, while 8,420 crime victims remained un-compensated.

One important reason is the backbone of the problem itself: crime. Due to the surge in violent crimes, victims now exceed the previous year by 22%.

In California, it takes an average of eight months before the State pays the claim to the victim.

This compensation law is a good one. It should be uniform throughout the nation, and the federal government should assist the states. Someday this may be a reality. Ask your Congressman and Senators to introduce a bill to compensate victims.

Our country should be responsible for the safety of its citizens. The weak laws imposed upon the criminals permit them to terrorize us in the streets and in our homes. The least we can ask from government is to provide some type of compensation to cover medical and related expenses incurred as a result of becoming a victim of a violent crime.

For citizens who care, there is something you can do to help catch the violent criminal.

"We Tip Hotline" For The Anonymous Caller

The movies of the early 40's depicted such superstars as George Raft and Edward G. Robinson, Jr. referring to an *informer* as a squealer, a dirty rat, a stool pigeon, a fink. Being an informer had a nasty stigma to it. Four decades later, the taste is the same—no one wants to get involved; no one wants to snitch on the criminal because many are afraid of reprisals.

The criminal strikes anytime, anyplace. His victims come from the ghettos, the middle class areas, and among the wealthy. Frequently, someone, not involved in crime, learns the identity of the criminal. How do they react? How should they react? We examine two actual case histories (only the names were changed).

* * *

Tony's Famous Hamburger Stand has been the hang out for local teenagers for over 30 years. They make the best chiliburger in town; greasy and good.

José sits on a corner stool. He finishes a bite of a chili pepper, looks over to his long time *compadre*, Frank, then quickly looks around to make sure no one is listening. He whispers, "Hey man, you know that dude that sells kids dope? That dude ought to be in the joint, man. He sold my kid sister some 'shit' yesterday. Blew my mind when I found out. Hell, she's only 13."

José's eyes danced around in his head. He took a bite of his burger, and continued. "Frank, I want to put him out of business, you know what I mean, man? She's my sister. But if I kill the bastard, they'd put me in jail, so that son-of-a-bitch'll stay on the streets selling his junk."

Frank doesn't know how to respond. He hesitates, wipes some chili off his face, and says, "Right. We don't need *this* guy in our neighborhood. But what the hell can *we* do about it?"

Clear across town, two other good buddies are having lunch. Not chiliburgers. Philip sits in a booth on the top floor of a restaurant in Century City. He sips a glass of wine and stares at his long time friend, Benjamin. Philip looks worried. While talking about the stock market, Philip suddenly switches to a personal tone.

"Ben, I haven't slept at all during the past few days. I have to talk to someone. I trust you. Have you heard on the news ..." The waiter approaches with their lunch, and Philip becomes silent, until the waiter leaves. "You know about that young girl who was raped and murdered—strangled? It happened last week in the shopping center?"

"Yeah," says Ben.

"I think I know—no—I *do* know who did it."

"Oh God."

They look into each others eyes. Phil's voice is tight, choked up. He says, "I received a phone call from my no-good brother-in-law." He pauses and takes a fast swallow of wine. "He told me he needed $5,000. Stake money, he said. To get out of town for awhile, that the bookies were after him. Ben, in all the years I've been married to his sister, I don't think he once told the truth. I didn't believe his story, and he knew it.

"Well, for the first time in his life, I think he really was on the level with me. He said he was in trouble. He confessed, 'I made a mistake.' Then he told me something I wish he hadn't. His excuse was that he was drunk, drunk out of his mind. He said he raped the girl in the back of his van. She screamed, and threatened to tell the police. He claimed he panicked, took his belt off, and, well you know the rest. It's like a nightmare.

"Reluctantly, I offered to pay for a lawyer, and I pleaded he turn himself in. He refused. That bastard kept reminding me I was married to his sister, and that we're family. Before I slammed the receiver down, I told him our family ties were finished. Ben, I don't know what to do. I just hope the police catch him."

Ben is wise—at least Philip believes so.

Spontaneously both men reach for the bottle of white wine and refill their empty glasses. Benjamin takes two quick swallows before he advises his friend, "It's probably best to leave this to the police; they'll eventually get him." This sounded reasonable to Philip. He wouldn't have to get

VIOLENT CRIMES

involved, there would be no problems in the marriage, and it would finally be off his mind.

They finish lunch, and say goodbye.

Has the situation been resolved? Both men approach their vehicles and have second thoughts; something didn't occur to them during lunch. What if he rapes again, or kills another victim? Now, Phil and Ben start to hear their consciences.

Back at Tony's Hamburgers, José finishes his second chiliburger and walks toward his car, thinking about the other kids who'll get hooked by the neighborhood dope dealer. He wonders, "Is there something I can do to stop it without getting involved?"

* * *

You aren't a squealer, a dirty rat, a stool pigeon, or a fink if you report the identity of a criminal. You will help *fight crime* and take criminals off the streets. One dope pusher will sell to thousands until stopped; a rapist will rape and rape, and may even murder. While "informant" might be a dirty word, if law-abiding citizens don't report information that can lead to the arrest and conviction of perpetrators of violent crimes, then crime will continue to increase, and you may well be the next victim.

WE TIP HOTLINE is a program that lets you anonymously assist law enforcement. This is how it works:

Dial toll free, 800-668-2800, to "tip" on any major crime: rape, murder, burglary, robbery, etc. Or call 800-472-7766 if it's related to arson. A pleasant voice answers, "This is your WE TIP line, do not give your name or identify yourself." The entire conversation is anonymous.

You're asked questions put together by law enforcement. The final question is, "Are you interested in a reward if the suspect is convicted?" If you want the reward—up to $500—you're given a code name and number. Your identity **is never known, not even to WE TIP. A reward is sent to the**

71

post office nearest you. The Postmaster only asks for your code name and number. You're paid in cash, and no identification is necessary.

The WE TIP program first began in California on February 15, 1972, and exclusively handled tips on drug pushers. On November 3, 1977, it was expanded to cover all major crimes. Today, fifty percent of all calls involve drugs.

On July 13, 1981, we interviewed Bill Brownell, former Deputy Sheriff, and Founder and Director of WE TIP. Here is their track record:

    22,238 tips received since 1972
    2,860 were arrested
    1,391 were convicted

This accounts for nearly 1,400 criminals taken off the streets, and about 10,000 less crimes that they would have committed.

In addition, WE TIP was responsible for recovering $45,500,000 in illegal drugs.

During the past three years alone they recovered $1,000,303 worth of stolen property.

In January, 1982, this California state-wide program will operate nationally. To find the toll-free number in your state, call 1-800-555-1212, and ask for WE TIP. Mr. Brownell told us the current hours "tips" are received are Monday through Friday from 7:00 A.M. to midnight and Saturdays from 9:00 A.M. to 5:00 P.M. The exact number of interviewers varies, but last year they logged over 350,000 work hours on the phones. The phone bill was over $7,000 per month.

WE TIP is funded by individuals and industry. Donations of any size are welcome and are tax free. Send your gift to WE TIP, INC., P.O. Box 740, Ontario, California 91762.

Soon, Quinn Martin Productions will produce a new television series based on WE TIP called, "Crime in America."

Well, José and Philip, now there *is* a way to deal with criminals—anonymously.

II
EVERY DAY SURVIVAL TIPS

The Obscene Phone Call

"Ohh ... You know what I want ... *don't hang up*—I need you, I love you. You turn me on—*don't hang up.*"

You want to hang up on the obscene phone caller, but somehow he compels you to listen.

"I know who you are ... and I want to get into your— "

That's it; you hang up. Seconds later, the phone rings again. You let it ring. And ring. And ring. Finally you pick it up. You don't say hello, you just listen:

"I'm warning you, don't hang up again, or I'll—"

You hang up. And now you've had it. You call the police, but you're told, "Nothing we can do." The police tell you to call the phone company. It's 10:30 P.M.; they're closed. Just after hanging up, the phone rings again.

This time you don't answer it. Throughout the night, it rings, and rings.

It's 9:00 A.M., and the phone company is open. You explain your problem to them, and their representative advises you to change your number. Adding insult to injury, the phone company insists you have to pay a fee for a number change, besides being victimized.

We won't even get into all of the inconveniences in notifying everyone of your new number.

* * *

You can't be physically injured over the telephone, but you can be mentally hurt by obscene or harassing calls. They can emotionally disturb and frighten you.

An estimated 25,000 such calls are made every day. There are federal and state laws prohibiting abusive, harassing, threatening, or obscene calls. Offenders are rarely convicted.

The phone company will only discontinue the caller's phone service. And they'll only do that if there's sufficient evidence of such a violation.

You have a telephone to call out and receive calls. When it rings you respond. Just like you do when the doorbell rings. You don't think about ignoring it, do you?

* * *

To hell with changing your telephone number, you decide.

Then you hear the sound which has become a nerve-racking nuisance—the phone ringing. "Guess who?" he says.

What do you do?

Well, you have a few options. Unfortunately very few. Changing your phone number is probably best if the harassment continues unbearably. Most persons who receive these kinds of calls are randomly dialed.

When the perpetrator behind the call sees he is capable of upsetting you, he'll continue to hound you. He *wants* you to engage him in conversation—he needs you to. So the best thing is to give him silence, followed immediately by a slam of your receiver.

Don't converse with him. Don't play with him. Don't tell him you're old, ugly, a cripple, or a policeman's wife. None of these really matter to the sick caller. He wants only your attention.

Don't talk to him; don't listen to him. And with luck, he might cross you off his list.

Blowing a loud whistle or "Shriek Alarm" into the mouth piece of your phone may deter him, too, but it's not always the solution, because he might just keep on harassing you.

What does Ma Bell tell us to do? Log the date and time of each call. Then, if the caller continues to harass you for several days, you can *demand* that the phone company put a "trap" on your phone. Usually they are reluctant to do this, mainly because it costs them time and money. If you can't get satisfaction from your telephone representative, then *insist* that you be connected with the *first* or *second level supervisor*. Request that a Special Agent of the phone company, criminal division, contact you immediately.

The "trap" is a method used to trace what phone number the harasser is using when you are called. The closer his location to yours, the faster he can be identified.

When caught, the violators are subject to fines, or imprisonment, or both, by federal and/or state authorities.

Dangerous and Deadly Toys

Consumers spend more than five billion dollars a year on toys. There are over 160,000 toy products sold in America. Every year over 625,000 children are injured playing with those toys. Some even die.

In 1969, the Child Protection and Toy Safety Act was signed into law. This was done to keep hazardous toys and other unsafe products for children off the store racks.

Not all toy-related injuries are prevented by this act. You must be vigilant in your toy purchases. Buy a toy appropriate for the age of the child, and don't go by the manufacturer's recommendations, because they often are wrong. So far there is no law restricting this labeling practice.

All electrical toys have federal warnings that the toy is not recommended for children under eight years old. Toys for infants should be sterilized to prevent the spread of bacteria.

Badly or poorly constructed toys, while they may be inexpensive, can cause many problems. Watch out for toys with sharp points and edges or brittle plastic. Toys containing small parts can easily be swallowed by small children. Small rattles are very dangerous. If it can fit into an infant's mouth, don't buy it. Toys that make loud noise, such as cap pistols, may cause permanent hearing loss if fired close to a child's ear.

Toy darts and arrows are not safe for small children. Some toys have a cord or string. If it is 10" or longer, don't give it to a small child, as it may wrap around their neck.

If the toy is packaged in a plastic bag, destroy the bag by cutting holes in it; a child can suffocate playing with a bag.

If you find a toy to be unsafe, contact the Consumer Product Safety Commission, toll free, (800) 638-8326.

Don't Advertise You Live Alone

Being married isn't any assurance that you won't be a target for rape or robbery, but living alone certainly adds to the risk. Criminals prey upon females living by themselves (or with small children) and the elderly living alone.

Before entering your apartment or condominium, you can be sure that to make it easier on himself, Bad Guy will check the names on the mailbox. *Your* mail box.

APARTMENT 3—BULLDOG AND MARY JONES; APARTMENT 5—ROBERT T. BROWN; APARTMENT 6—JOSEPH WHITE; APARTMENT

8—(your apartment)—PEGGY SMITH; APARTMENT 11—SGT. W.E. LOUIS; APARTMENT 12—MR. AND MRS. ANDERSON.

Guess which one Bad Guy's going to pick?

Apartment #8, your apartment. Why? Because you advertise that you live alone, and that you're female.

You could have fooled the would-be rapist or burglar if you had listed your name on the mail box as P. Smith or Smith family, or even Mr. and Mrs. Smith. Your neighbors may look at you strangely, but Bad Guy won't look your way.

The same goes for your telephone listing. If you can, request your telephone number to be unlisted. If listing your number is essential, then ask the phone company to omit listing your address. And when someone whom you don't know calls, never give it out.

Growl....

The growl of a dog scares most everyone. Talk to your mailman or gas meter reader sometime. When the canine is behind a closed door, inside the security of your home, and hears a stranger, most dogs instinctively bark or growl.

For the person or persons outside, hearing the sounds of the animal might force them to be a little leery of entering your home.

Whether the dog be a Doberman pinscher, a German shepherd, a toy poodle or a Lhasa apso, the sound of a dog behind your closed doors, to the ears of a would-be intruder, is a sound that need not always be tested for its potential trouble.

Here's a true story. A man and his wife returned home recently, and discovered their watchdog choking. They brought the dog to their vet for emergency attention. The vet told the couple not to go back home and showed them what had caused the dog's injury. Three human fingers were in his throat.

The police arrived at the house and discovered a burglar bleeding to death in the closet. He was too frightened to come out.

Criminals, unless insane, would rather pass by your house and look for another that doesn't appear to have a dog. Dogs offer protection. Owning a specially trained attack dog is one of the best protections you can have. But there's a drawback. Some attack dogs can be friend and foe to their masters, and may be a potential danger to children of the family. If you decide to purchase a dog, discuss the pros and cons with a recommended animal trainer.

Safety At Home For Seniors

Accidents are the third leading cause of death for senior citizens. Each year one million seniors die or suffer disabling injuries caused by accidents, and most of them happen at home.

You don't have to be vulnerable to frequent home accidents if you follow these safety tips.

You want your floors to look nice and shiny, but if they are highly polished, you can easily slip and fall. Throw rugs or scatter rugs are easy to slip on. A piece of furniture or a heavy object placed on part of the rug will reduce the probability of a fall.

It's not necessary to be physically handicapped to have grab bars in your bath tub or shower, and by the toilet seat. They assure you safety, and offer "a helping hand." A handyman can install these aids; he can also install an inexpensive help-alarm. A sound alarm of almost any type can be placed outside your door, wired to buttons, one in the bathroom and one by your bed, in the event you need medical or other help. Inform your neighbors to listen for the signal.

Your balance may not be what it used to be. Place non-slip mats in the tub or shower. Have an area rug, or even better, a rubber mat by the bath or shower. Keep a night light on in the bathroom; they plug into an electrical outlet, stay on 24 hours a day, and draw a minimal amount of electricity.

Unless you have small children staying with you, have your pharmacist put your prescriptions in regular—not child-proof—containers that are not difficult to open. Most medications have an expiration date on them. Check the date before using it. Some medicines lose part or all of their potency after it expires.

If you live in an apartment or condominium, or anywhere you have concerned neighbors, get together and form a watch-alert. If one of the seniors has not been seen for 24 hours, try to make contact. If you feel you can trust him, arrange for someone nearby to have a key in the event he has to enter your residence when you don't respond. When leaving home for overnight, let him know when you plan to return, and where to reach you if necessary.

And here's something more you should know ...

Special Warning For Senior Citizens

You're especially vulnerable to purse-snatching and fraud. Criminals prey upon the elderly. If, like most senior citizens, you're on a fixed income, loss of any amount of money can be disasterous.

Carry only small amounts of money in your purse or wallet. Pay most of your purchases with a check. Many banks offer senior citizens a free checking plan.

Of the over 41,000,000 crimes committed each year, almost half of these criminal acts were against one's property. Purse-snatching is one of the easiest of all such crimes. The national average "take" for a purse snatcher is $98; the pick-pocket theft averages $124.

To help prevent becoming a victim, hold your purse tightly, close to your body, under your armpit. Men, carry your wallet in a front pocket or button your hip pocket.

Each year, many victims attacked during a purse-snatching are seriously injured or killed. If an attacker grabs for your purse or wallet, and you're not immediately prepared to defend yourself with tear gas, let go; it's not worth your personal safety, regardless of the amount of loss involved.

Many criminals rob mailboxes, particularly during the mailing periods for social security checks (or retirement checks). Mailbox theft can be avoided. Ask your bank about the Treasury Department's Direct Deposit Program.

There is no cost to you. The check is sent to your bank, and deposited into your checking or savings account. You avoid the long walk to the bank or the postage cost to mail the check.

Throughout this book you'll find many tips to avoid fraud. Follow them. Stay away from any "get rich quick offer." Don't let anyone convince you to part with your money or life savings—you may never see it again.

The Deadly "Loaded" Weapon

When Have You Had Too Much To Drink To Drive?

Greg felt the effect of the last beer he just downed. Leonard, the quarterback of the football team at Greg's high school, pressed another ice cold beer against his arm for Greg to take.

"Uh ... no, I don't think so, Leonard," Greg says, feeling the eyes of the rest of the team, lounging in one of the teammates' garages, coming upon him. "I've had four. If I

EVERYDAY SURVIVAL TIPS

have another, I can't drive."

"That's crap, man. I've had five, and I can drive."

"Hey, Leonard, you're a poet."

They all laugh. Greg doesn't like the pressure. He'd prefer to be accepted. Being the smallest guy on the team is hard enough; he constantly has to prove himself to the rest of the team, and now that they have invited him to Paul's garage "to party," he doesn't want to blow his chances at really feeling a part of the team—on and off the field.

Paul says, "Come on, Greg, really. One more won't hurt. Can it?"

Greg looks at the beer, then at Leonard.

"We all drink about a six-pack each," Leonard pipes up, "and we got no problem driving anywhere."

"I drive better drunk than sober," says a teammate, who looks to Greg like he could drink two 6-packs and never feel it due to his incredible size—over 225 pounds.

"You guys are a lot bigger than I am," Greg points out.

"So?"

"Well, the alcohol in you can be in greater quantities than the alcohol in me."

"Listen to that. Who told you that crap?"

"It's true. The heavier the person, the more he can drink and be less affected by it."

"I think your mommy and daddy have you brainwashed," Pete says.

"No," Greg replies. "That's medical fact."

* * *

Eight thousand young Americans are killed each year in accidents involving alcohol. And 40,000 more are disfigured and crippled, and hundreds of thousands are injured.

Your sons and daughters can be victims of an accident even if they aren't drinking excessively. The person behind the wheel can be the one that will cause the serious accident, and your child, riding along in the passenger's seat or the rear seat, can be seriously or fatally injured.

How do you and your children know if a driver has had more to drink than safe driving allows? Well, Greg is right. Here is the guideline: based on a two hour period, the amount of drinks (1½ ozs. of liquor or 12 ozs. of beer) consumed in reference to the person's weight will tell you if the person has had too much.

PERSON'S WEIGHT	DRINKS IN TWO-HOUR PERIOD 1½ OZS. OF LIQUOR or 12 OZS. OF BEER
100 lbs.	2½ drinks—*driving impaired*. 3 drinks—*don't drive*.
140-160 lbs.	3 to 4 drinks—*driving impaired*. 5 or more drinks—*don't drive*.
180 lbs.	4 to 5 drinks—*driving impaired*. 6 or more drinks—*don't drive*.
240 lbs.	5 to 6 drinks—*driving impaired*. 7 or more drinks—*don't drive*.

When driving is impaired, there will be .05 to .09% of blood alcohol, and the driver may be legally drunk. And in any event, he is a candidate for a serious accident.

The "don't drive" limits will show a .10% blood alcohol or more. This driver offers even greater risk to his passengers, himself, and others on the road.

>Myth: A can of beer is less intoxicating than an average drink of liquor.
>Fact: A 12 ounce can of beer, 1 and a half ounces of liquor, or 6 ounces of wine are equal in effect on the body.
>Myth: Black coffee or a cold shower will sober a drunk.
>Fact: Alcohol is eliminated from the blood stream only by action of the liver. The process takes time, not coffee or cold showers.

* * *

Greg pictures himself in his mangled Toyota. He sees himself bleeding profusely, and feels the pain and agony of serious injury. The beer the team is offering him just doesn't seem worth the price for peer acceptance. His father was killed in an accident only three years ago. He was hit by a drunk driver.

Greg tells them all, "I'm sorry guys, but just like I explained, I'm at my weight's limit. I have to drive home tonight, and frankly I want to make it home without smashing my car or smashing me. I want to hang out with you guys—you're a lot of fun and you're all friendly, but if joining the gang means testing fate, well, I'd prefer not to be a member."

The team is somber. He's being brutally honest. And seeing he's only one small fry against them all, they can't help but respect his ability to say "no."

"Forget it, Greg," Leonard says. "You've had enough." And he playfully gives Greg's shoulder a punch.

How Do You Protect Yourself From The Drunk Driver?

Following a few social drinks, Mr. Nice Guy, who happens to be your next-door neighbor, gets behind the wheel of his car. The street divider looks wavy; as a matter of fact, he sees two white lines instead of one.

Crash ... head on. Mr. Nice Guy, who is not a criminal, walks out of the car without a scratch, but a mother and her child are dead.

Nearly 26,000 vehicle fatalities occur in America each year because one or more of the drivers were drunk. That's 250,000 lives in the past ten years—more casualties than both sides lost in fighting in Viet Nam. Besides deaths, drunk drivers are responsible for 1 million crippling injuries each year.

HOW TO PROTECT YOUR LIFE AND PROPERTY

The "loaded" driver, whose vehicle becomes a deadly weapon, is generally thought of as a fun-loving person who just had a few too many and not as a criminal.

How do you protect yourself from the drunk driver? First, *you* must remain sober. When two drunks meet in the streets, the odds double that one of the two will hit the other.

Drive evasively. Watch your mirrors. Watch out for the other guy. Stay alert; don't let yourself become distracted by the radio or passengers. If a vehicle is traveling well below the speed limit, he may be drunk, so stay out of his way.

If a weaving car is just ahead of you, slow down, or pull off the road for a few moments, because there is a good chance he's about to cause an accident, and you don't want to be behind him.

Perhaps one of the most dangerous times to be on the road is shortly after the bars close. You can be sure that many of the people leaving the local pubs across America have had too much to drink, and for too many of them, it's the start of the often tragic street Demolition Derby.

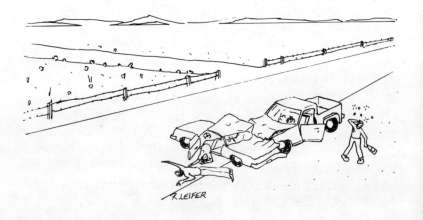

Does Your Youngster "Do Drugs"?

What was considered morally wrong in society not long ago is fairly well accepted today. "Shacking-up," living together not as man and wife, a decade ago, would have shocked the parents of both couples, but today, this way of life is acceptable to many.

This is the 1980's, and things have drastically changed. Smoking marijuana was once a strong social taboo. Today, it's almost legal, and its users include perhaps 10% of *Who's Who in America.*

When you smoke a cigarette you are violating your body; you become a candidate for cancer; you may shorten your lifetime.

If you drink alcohol excessively, you also endanger your body, and may die sooner than necessary. Parents who smoke and/or drink, and condemn others who use marijuana, frequently are rebutted by these facts.

Who's really right?

For a long time, critics of drug use argued that it starts with pills, then you graduate to marijuana, soon you resort to cocaine, morphine, LSD, PCP, and finally you're hooked. Not everyone who smokes pot (marijuana), does coke (cocaine), or other drugs. Our concern is a youngster's experimenting with any type of drug or narcotic. Drug use and the child's immaturity often result in drug abuse.

If your youngster is using pot or drugs, it may be difficult for you to detect. Some have obvious symptoms, some don't.

Physical evidence that may convince you of your child's indulgence with the "weed": if you find among his/her property a pot pipe, a small pipe, or papers to roll marijuana (the type used years ago to "roll-your-own"), which are openly and legally sold almost anywhere, including your neighborhood grocery store, you can reasonably believe the youngster is smoking "grass."

When paraphernalia is found (widely sold in record

HOW TO PROTECT YOUR LIFE AND PROPERTY

stores) youngsters will often bend the truth and say it belongs to a friend. A classic symptom of a drug-abused youngster is red eyes. But non-prescription eye drops often erase this symptom.

If you suspect your youngster is "stoned" on drugs, put him/her to the test. Have them click their fingers after waiting for one minute of time to pass. If they're on drugs, most likely the finger clicking will start within 10 to 20 seconds. Why? They'll have a distorted sense of time.

A youngster using drugs often thinks he is driving very slow, when in fact he is actually speeding.

Other signs to watch for:

If the youngster loses interest in school, or other activities, particularly ones that were very important to him, he may be on drugs.

A drug user frequently uses incense to hide the odor of "pot."

When addicted, the appetite diminishes, but there's a strong craving for sweets.

The youngster's mood changes.

His memory may be impaired.

A chronic cough or pain in the chest may be present.

Often, there is depression, and a feeling of loneliness.

With a female, there can be irregular menstruation.

Drug abuse can cause a disturbance in sleeping habits.

A user often pays less attention to grooming and cleanliness.

Common infections become more difficult to resist.

Strange or secretive phone calls may be made or received.

Money or other valuables may "disappear" from the household.

If your youngster has one or more of these warning signs, there is a possibility he or she is doing drugs.

III
DISASTER, EMERGENCY & MEDICAL PROTECTION

We have to be prepared. Although there is little we can do to prevent a natural disaster, there is much we can do when we're faced with one.

Earthquakes

If you live or visit in California or other earthquake-prone states, expect to feel the shake, rattle and roll of one someday. If the "big one" strikes (many California seismologists predict it to be an eight on the Richter Scale) you could be in grave danger.

Before the earthquake occurs, keep the following in your home:
1. A *first aid kit*.
2. *Flashlights* with extra batteries —and *candles* and *matches*.
3. One week's supply of *canned foods* and *powdered milk*.
4. Three gallons of *fresh water* for each person in the family.
5. A *portable radio* with extra batteries. If you own a *Police Emergency Radio scanner*, you'll be able to tune in to local emergency bands.

HOW TO PROTECT YOUR LIFE AND PROPERTY

6. *Pipe* and *crescent wrenches* to turn your gas and water off.
7. Determine the location of the electric, gas and water *main shutoffs*. The white section of some telephone books offers instructions on how to shut off your gas and electricity. If not, call your utilities and ask for the procedures.
8. Don't keep any *heavy objects* over beds.

When the quake hits, don't panic. Stay calm. If you're indoors, stay away from the windows, masonry walls, and chimneys. Don't go outside, because you put yourself in greater danger from falling objects. Get under a bed or under any heavy piece of furniture, like a table or desk. Or stand beneath a strong doorway or against an inside wall.

Now, if you're caught outside during the trembler, stay in the open, away from utility wires, buildings, trees, and anything that could fall on you.

When you're driving your car, you might not feel a moderate quake due to the car's motion. If you feel it, or if your radio alerts you to one occurring, just pull over, staying in an open area away from structures, utility poles, overpasses, et cetera. *Don't get out of your car.*

Aftershocks may follow a big earthquake. As soon as you feel safe, check for fires and fire hazards. Open the windows if you smell gas, then shut off the main valve. Leave, report the leak to the gas company, and don't return until they check it.

It may be best to shut off the water main and the electricity until you have an opportunity to see if the water pipes were damaged or if there is damage to electrical wiring.

Flood

Unlike earthquakes, there is normally a warning when a flood is expected. If you have a radio which broadcasts the weather reports 24 hours a day, stay tuned to that during severe inclement weather. You can also listen to your local radio and television for weather reports.

Before the flood, fill your vehicle's fuel tank. Stock food requiring no refrigeration. Fill clean bath tubs with fresh water. Have a portable radio, flashlights with extra batteries, and keep both candles and matches handy.

When you're instructed to evacuate, do it immediately. If there's time, turn off your gas, electricity and water, and take your valuables to a higher level.

As soon as you receive a flash flood warning, go to higher ground. Walk if your car stalls. While driving, stay alert for slides, electrical wires dangling across the road, and road damage in general. Don't drive through flood water; it can be much deeper than it looks. While walking, don't cross areas where the water is above your knees. The reason for this is: under that murky water could be holes, cracks, sharp objects, and other invisible dangers.

Lightning

They say lightning never strikes twice. Great. But all it takes is once. When you're indoors during a thunderstorm, deprive your curiosity by staying away from windows and doors, fireplaces, stoves, sinks, pipes, and electrical appliances. Unplug your television set's antenna wires. We know that's tough to have to do, but try listening to your portable radio, or, better yet, read a book by candlelight. Don't use the bath tub or shower, your electrical appliances, or your telephone, unless you want to gamble and get a good shock.

If you're outdoors, and you can't get into a building or other structure, or into a car, stay away from high ground. Well, you might be thinking to yourself, what if there is a flash flood—where high ground is recommended—what then? Well, the flash flood is probably a sure thing, whereas being hit by lightning is not. We suggest, in this case, to go to higher ground.

While outdoors, you shouldn't touch metal objects, or get into water. And stay off small boats.

Lightning may be about to hit you if your skin tingles or your hair stands on end. If this happens, immediately drop to the ground. But don't confuse these sensations with static electricity in the air during especially windy periods, or your dropping to the ground might become an embarrassment.

House Fires

If you're indoors, get everyone outside at the first sign of smoke or fire, *closing the door behind you*. When you're safe, call the fire department. Many people die in fires because they do the opposite: they call the fire department, then evacuate.

If you're not fortunate, or prepared enough to escape, and you're trapped in a room, don't panic. *Stay near the floor*. If you can't escape out the outside window, at least open the window at the top so the heat and smoke will go out, and open the bottom to permit the fresh air to come in, but *only* if there is no smoke outside.

Check all the doors, using your hand, before opening them to escape. If it's hot, don't open the door. If the door is cool, *slowly* open it and stay behind it; then if you feel heat or any pressure coming at you through the doorway, immediately slam the door shut.

Remember: If you must escape in a smoke-filled area,

always stay near the floor where the air is better. If you can, cover your face with a wet hankerchief or any available cloth.

What do you do if the clothes you're wearing catch fire? No, you don't scream your head off. You quickly cross your arms over your chest, touching your hands to your shoulders. Drop to the floor and roll over slowly until the fire is out. If there is a rug, a coat, a blanket, some drapes, anything, wrap yourself in it and roll on the floor. When the fire is out, don't pull the clothing from your skin, or you'll find yourself painfully peeling like a naval orange.

What You Need To Know About Fire Extinguishers

Fire strikes more than one-half million homes each year. For the average person, any fire extinguisher has but one purpose: to put out a fire.

Wrong.

Like the good doctor who specializes in a particular form of medicine, most home fire extinguishers specialize in putting out special fires.

Do *you* have the correct one at home?

To find out, look at the label or decal on the fire extinguisher. If it's marked CLASS A, this will put out a fire involving wood, many plastics, paper, cloth or other similar substances.

CLASS B puts out fires involving flammable liquids or grease.

CLASS C is for fires involving electrical equipment.

You can own all three. But that would be both costly and take up a lot of space. Since you have no way of knowing which type of home (or office) fire you will face, the solution is to purchase one extinguisher that is listed CLASS

ABC. It contains a dry chemical concoction that works well on all three major types of fires. The average cost of a small 5-pound size is about $25 or less. They are rechargeable, and should be serviced once a year.

Read the instructions BEFORE THE FIRE so you are familiar with its operation, and keep the extinguisher centralized within the home.

Each member of your family should be instructed in the use of the fire extinguisher. If you have other questions about your fire extinguisher, you may call your local fire department, and they should answer your questions.

How To Deal With Fires

Unless deliberately set, most household fires are either accidentally set or they start because of some faulty electrical problem. Unfortunately, most fires occur when you are sleeping, between midnight and 5 A.M..

When a fire starts, usually you are sound asleep, so it's very difficult to think and act clearly during those most crucial first few minutes.

Thanks to smoke and heat alarms, even you sound sleepers can be awakened when a fire is threatening your life. It is highly recommended (and in some cities, mandatory) to have smoke detectors in every dwelling.

Most alarms of this type cost around $10-15 and are the best "life insurance" money can buy.

Kitchen Grease Fires

If you can reach the stove, turn off the flame at once. Never attempt to extinguish a grease type fire with water. This type of fire won't be stopped using water. Instead, water creates a scalding steam and injures anyone near it.

Use only a foam type fire extinguisher, or class ABC, or, throw a lot of salt on the fire. If the fire is in a pot or pan, put a lid on it; do the same if the fire is in an oven.

Electrical Fires

If a fire breaks out in a wall, immediately call the fire department. For all electrical type fires, shut off the electrical source by tripping the circuit breaker, on the main switch, to *off* . If you have a fuse, pull it out.

Fire Drills

Fire drills were routine in school. If a real fire occurred, we knew just what to do. But what will happen if a fire breaks out in your home? Is the family prepared? What about youngsters? Can they cope with a fire emergency at home? Chances are they will panic; perhaps you will too. This is normal, and it's also deadly. Periodically at home, you and your family should conduct a fire drill. Show and practice where to escape through normal exits and doors, windows, etc. Nearly everyday, many children and adults are found dead because they hid under a bed, or in a closet during a fire. Most physicians recommend preventative medicine. So should you also consider preventative fire methods. Purchase fire extinguishers, and place them strategically throughout the house, and periodically check to make sure they are properly charged.

All family members should sleep with their doors closed. In the event of a real fire, this will keep noxious fumes out of the bedroom.

Fire itself is not the major killer. It's the fumes that kill most victims.

So have a plan ready, a plan that'll allow all members of the family to escape, and meet somewhere outside the house.

In summary, if you follow these suggestions and basic

procedures, there's no guarantee you won't lose property, or won't be injured, but it may save your life or the life of someone dear to you. Just remember: You and your loved ones are irreplaceable, so preventative measures can only help in the event of a natural disaster. Naturally.

How To Survive A Hotel Or Apartment Fire

The deadly fire that swept through the MGM Grand Hotel on November 21, 1980, injured 500 and killed 84 persons. It was a terrible disaster, but it served one valuable function: it made us all more alert to the possible danger, yet, few of you know exactly what you must do to survive.

In many better hotels and apartments, the top floors cost more because you have the best view. In the event of a fire, the occupants above the seventh floor have a lesser chance of survival. Today's space age fire departments, in even the largest cities, don't have fire trucks capable of reaching beyond the seventh floor.

What is worth more to you? The view or the safety? We wish everyone of our readers would follow this advice every time they check into a hotel:

First thing you should do when checking into your room is not to call room service but go out into the hallway and look for the fire exits on your floor. Be sure each member of your party is aware of the locations. That still isn't enough. In the event of a real fire, will the fire exit open? Find out, test it. If it won't open, immediately call the hotel manager or security, and have them open the exit before you retire for the night.

In most cities, hotels are not required by fire code to have fire alarms or smoke detectors, particularly in older buildings. Today, alarms of this type are small and rather inexpensive. If you travel frequently, you may want to buy one for this purpose. If you are going on vacation, and no one is home, take the home alarm with you.

If your detector alerts you, or if you don't have one and you smell smoke, call the fire department *first*. If you call the hotel, they will usually send someone up to check it, and that will waste valuable time and perhaps lives. Tell the fire department the name of the hotel and location. Remember, though, hotel chains have more than one hotel in the same city. Be sure to give them your location and room number.

Then you can call the hotel for assistance.

After you summon for help, put your hand on the doorknob. If it feels hot, DO NOT open the door. In the event the doorknob IS NOT HOT, open the door, take a quick look into the hallway. Yell FIRE!—FIRE! Unless there is a lot of smoke, you may attempt to escape to the fire exit (you previously checked it and it opens). Before going into the hall, place a wet towel over your face, allowing just enough opening to see.

If the fire is small, like if a couch in the hallway is on fire, and you see a fire extinguisher, you may want to put it out yourself. Be prepared, though; there may be other areas on fire. Continue to alert other occupants by yelling FIRE! DO NOT attempt to use the elevators. If there is a fire, it can short the electrical power and possibly strand you inside the elevator, which could be disastrous.

We are always advised "Do Not Panic" when we personally face a possible life and death situation. How can we not panic? It's human nature. Try—try real hard. Stay calm if you can, and think only of how to escape. When you go down the fire escape, *don't run*. You may fall and be injured or killed. Hold onto the rail, unless it's hot. If it's real smokey, hold one hand of each member of your group and keep together. If a lot of smoke is in the area, you may decide to climb up the fire escape, instead of down, unless you are nearing the last floor and feel you can safely make it down. If you do retreat to the roof, there is a good chance firemen are already up there to help you to safety.

What do you do if you are forced to remain inside your room? Again, try to remain calm. Do not break the window. Sure, you've seen many movies when the hero did just that, but doing so will jeopardize your safety. If there is smoke outside, you want to keep that smoke out of your room, so the windows must be closed.

Fill the bath tub and sink with water. Wet all the towels, sheets, blankets and pillow cases. Stuff them around the door or anyplace where smoke can enter the room. Keep them wet. Block all vents with wet objects. In the event of a gas leak, don't smoke.

If the walls are hot (touch them with your open hand) use an ice bucket, or anything that will hold water, to throw as much cool water on the walls as possible.

Many persons don't know that fire isn't the major killer during an actual fire; it's smoke inhalation. Wet towels placed over your nose and mouth can prevent the killing smoke from being inhaled.

Other important tips in the event you're trapped in a fire of any type are pointed out elsewhere in this book.

The main thing is, don't give up—fight back. Professional firemen will soon arrive.

The police receive hundreds of calls in major cities each hour, and have only a limited number of police officers to respond, but fire doesn't occur every second as crime does, so the fire department and ambulance rescue can respond at once. In most cities, they must leave the fire station no later than 60 seconds after the call is received.

Household Emergencies And Safety Precautions

Power Failure

If your electrical power fails throughout the house, it is

usually a result of a power outage caused by some breakdown in your neighborhood. Look outside. If everyone's lights are out, be patient; the power company will restore them as soon as possible. But always have a flashlight and candles available in the event you have a power shortage.

If your neighbor's power is working, the trouble may be in your electrical system. Older homes may use fuses. Look at your fuse box, and if you find one with a blackened window, or the silver strip behind the window is broken, your fuse is dead. Always keep spares and replace it.

Most modern residences have circuit breakers. Sometimes a circuit breaker becomes overloaded, and the circuit shuts itself off. This can be corrected easily by flicking each circuit breaker from the *on* position to the *off* position, and then flick it back to *on*.

If your lights dim or flicker when an appliance is turned on, or the circuit breakers trip or fuses blow often, or the television picture shrinks or acts up when an appliance is plugged in, or it affects the radio, this is an indication that your house is not adequately wired, and you should contact a licensed electrician to correct the problem.

Appliance Smokes Or Appears To Be Burning

This often happens. If your appliance starts smoking (and we don't mean a piece of toast is burning because it was set too high), or there is an odor of fire, unplug the appliance, but don't touch it, just the power cord.

If the plug is on fire, go to your main power box and turn it off.

If the appliance is on fire, extinguish it with a Dry Chemical Extinguisher, or smother it with blankets, a rug or other such object.

Downed Power Line

A power line sometimes falls because of heavy snow or

ice. If it's struck during a traffic accident, the power line may also fall. Never go near a line that is down; call the police and department of water and power immediately. You should also set up flares or a barricade (chairs or other objects) to warn passersby.

Plumbing Emergencies

If a pipe leaks, turn off the water supply. There are separate shut off valves under sinks, bath tub panels, toilets, etc. If you don't know where the source of the leak or break is, turn off the main water shut off valve. Learn where it's located in advance of an emergency; your department of water and power can help you if you have trouble.

After the water is off, you shouldn't try determining the extent of the damage until you wipe the area around the leak dry. Don't wait to repair even a minor leak, or you may find yourself with more damage, because leaks will usually enlarge with time. Water damage is disastrous. Leaking water can ruin plaster walls, wallpaper, it can stain your paint, and, most importantly, develop an electrical hazard.

Since leaks usually start at threaded joints, you may be able to fix a leak by simply tightening the fitting. A warning, though. Deteriorated pipe often breaks while you're tightening the fitting; if it does, epoxy compound should help solve the problem. Even after repairing small leaks yourself with electrical tape, you should consider this only temporary. Call a plumber and have it repaired properly.

Frozen pipes can be dangerous. There must be an escape passage for water and steam. Heat should be put into the area where the pipes are frozen to restore the flow. If the area cannot be heated, such as a crawl space or cellar, you can use a wet proof heating pad, or a propane torch, household iron heating cable, or you can simply wrap rags around the pipe and pour boiling water over them. Whichever method you decide to use, you should always open the nearest faucet on the line that is frozen and work back from the opening to the frozen section of pipe. If you don't, the buildup of pressure can be dangerous; there must be an escape for the water and steam. *Caution: Do not heat a pipe so hot you can't hold it. An overheated pipe can produce enough steam pressure to cause an explosion.* When using a propane torch, use a flame spreader nozzle and move the flame back and forth to prevent hot spots from developing; place an asbestos sheet behind the pipe. If the pipe has split, leave it alone. Wait until you have the proper repair materials.

Sometimes a clogged drain can cause severe water damage. If the drain is clogged, pull the stopper out and look for hair and other foreign objects. You can use a plunger, and if that doesn't do it, a snake (purchased in most hardware and plumbing retail outlets) or a strong chemical solvent will usually do the trick. If it still won't clear, call a plumber.

If you lose your ring, or other valuables down the sink drain, *don't run the water.* Remove the sink trap, then simply take out the valuable item from the trap.

Appliance Shock

Electric blankets, coffee makers, fans, curling irons, stylers, hot plates, irons, toasters, vacuums and any other electrical appliance, can shock you.

Leaking current because of faulty insulation, a frayed power cord, or a component that has touched metal while being electically active will cause shocks.

1. *A short circuit in the power cord of your appliance:* Disconnect the cord. If it's built-in, detach the leads. You must then go to your local electrical supply dealer and purchase a volt-ohm meter. It will run between $11 and $17, but it is definitely worth it. When you

get it home, clip the probes to each plug terminal. During the test, pull and twist the cord; this will aid your test in determining a short circuit. (Receiving a shock from the power cord usually means it needs repair.)
2. *Current leakage*:
Unplug the appliance. Clip one volt-ohm meter probe to the body of the appliance. If the cord is separate, clip the other probe to one of the terminal pins of the appliance, and, if not separate, clip it to the plug terminal. (Receiving a shock while you touch exposed metal of the appliance usually means either a grounded wire or current leakage.)
3. *Defective internal wiring*:
Set up the volt-ohm meter test as in number two above. Open the appliance housing. Now the wiring is approachable, so gently tug the wires, one at a time.

One last tip: when you go buy your volt-ohm meter, you may look around and find a volt and amp meter for about $50-60. Any meter you do buy should be demonstrated for you before leaving the store.

Trash Compactors

This large appliance helps dispose of waste. The waste is crushed into tight bundles that are easy to carry or store in trash cans. The power screws in the compactor ram downward to a force of about 2,000 pounds before the motor automatically stalls, reverses the direction of the ram, and lifts away from the compacted waste.

Manufacturers have installed safety devices and made the disposer difficult to disassemble. Here are some rules to follow to avoid problems:

1. Don't put aerosol cans or bottles containing flammable liquids, strong chemicals, or insecticides, or cans into the compactor.

2. Don't put in glass bottles standing straight up. Always lay them flat.

DISASTER, EMERGENCY & MEDICAL PROTECTION

3. When you empty the compactor, you want to be sure to avoid being cut by glass that has pierced the sides of the bag. The compactor bag should be held only from the top.

Small Engines

Periodic maintenance and repair of your small engines (like the one on your power lawnmower) can help prevent mechanical problems. A clean engine usually is trouble-free.

But there are, of course, some safety precautions before you begin work.

Disconnect the spark plug cable before you work on an engine or the engine might start on its own and injure you. The cable shouldn't dangle free, though; if there's a rubber boot, pull it back and expose the connector. Then attach it securely to an engine cooling fin or, if there is one, the special grounding clip.

If the engine is placed on its side, make sure the oil fill hold and gas cap are facing up. Spilled gas or oil can be very dangerous. And if you refuel the engine, move the engine at least twelve feet away from the refueling spot. If you don't, fumes and exposed fuel can ignite.

This may be obvious, but every year, do-it-yourselfers burn themselves on hot engines. Let your engine cool before you work on it. Even a hot engine can ignite fuel, and if you spill some on the engine when it hasn't cooled enough, you could be seriously burned.

Chain Saws

We all like to save money. Tree trimmers, gardeners, and handymen are expensive these days. So we suddenly find an array of small, home-use chain saws on the market. Chain saws are manufactured for all types of cutting work, heavy to light. If you plan on easing your pocketbook's burden and get a tan working in your backyard forest, keep these safety precautions in mind *all the time*.

Start the chain saw before you allow it to touch the wood. If you forget, the saw could kick back out of control.

Hundreds are injured by not making sure that the *saw is not obstructed in any way by surrounding tree limbs, rocks, or the earth.*

As with small engines, after refueling, *move the saw away from the refueling area before you start it.*

Wear work gloves when using the saw; it's also advisable to *wear earplugs* if the sawing is for long periods.

Don't use the saw in semi-darkness. If you're tired, ill, intoxicated, or under the influence of drugs, your coordination is impaired, so stay away from your chain saw.

Lastly, never allow young children to use a chain saw.

Lawn Mowers

While mowing that lovely lawn, always wear heavy shoes to protect your feet—steel-tipped safety shoes are best.

Before you get buzzing across the lawn, though, pick up anything that might get under the blade and be flung from the mower to cause damage or injury: toys, rocks, sticks, and other objects.

A good friend of one of the authors lives in the foothills of Los Angeles. He has a sloping lawn, and one day while mowing *across the slope,* the mower side-slipped on the dewy grass, the wheels caught, and over the mower went, exposing the whirling blades. Take heed: never mow across; always mow *uphill and downhill* to prevent the mower from flipping over.

Pesticides

Some people tend to feel safe when administering pesticides around the home. They think if it's made available for the general public's use it must be safe. Not true, not true, not true. You might get by with using some things without reading the instructions, but not pesticides. They can be dangerous if the instructions are not understood. Read the label carefully.

How do you read a pesticide label?

The law requires pesticide labels to display certain information: brand name; whether it's liquid, powder or aerosol; its uses; active ingredients; amount in container; directions for use; Environmental Protection Agency number; and an indication of the pesticide's hazard effects upon humans and wildlife. "Caution" means it's *relatively safe* to use; "Warning" means it's *moderately toxic,* and "Danger" tells you it's *highly toxic.*

Here are some other precautions:

1. Keep pesticides away from food, eating and cooking utensils and areas for cooking.

2. Don't inhale it or get it in your eyes or on your skin.

3. After using pesticides, wash your hands and face thoroughly.

4. Never use pesticide near an open flame (a pilot light or furnace), and put out your cigarette, too, while you're at it.

5. Leave the room after using aerosol spray, and close the doors behind you. Stay out of the room for at least thirty minutes, longer if the label tells you to.

6. Ventilate the room after the recommended time.

7. A note about those pest strips: if people are going to be in the room for prolonged periods (a party), especially if elderly persons or infants are present, don't use them. And keep them away from areas where food is cooked and prepared.

8. Store pesticides in a cool dry place, out of the reach of children and animals—never near food.

9. Pesticides should never be disposed of in a way that could contaminate water or wildlife; don't flush down toilets, sinks or sewers.

10. If you swallow pesticide, follow the antidote instructions on the label and call a doctor immediately. Be sure to bring the container to the doctor; the Ortho Division

HOW TO PROTECT YOUR LIFE AND PROPERTY

of Chevron Chemicals has a national emergency number for use by physicians: (415) 233-3737.

Household Insect Control:

Ant: Most ants nest outside in burrows or mounds, under concrete slabs, cement crevices, or in dead wood. Some nest indoors under flooring, inside walls, basement, garage, under papers, or in remote corners of the house.
To control ants, follow them to their entrance and exit hole. Cover the area with an insecticide containing chlordane, dieldrin, diazinon, lindane, or malathion. If you find the nest, treat it with insecticide dust.
Cockroach: There are five species of this house pest, ranging in size from ½ to 1 inch in length and from yellowish or reddish brown to black in color. The adult German cockroach is light brown with black stripes, about 5/8 inches long, and has a resistance to many pesticides. All species tend to be active at ground. Between 12 hours and three days, maggots hatch out.
Fly: To control house flies, maintain screening and good sanitation. Screens should have at least 14 meshes to the inch. Cans should be kept covered; seal garbage bags. Don't leave foods exposed. Your pets' excrement should be picked up every day and disposed of as garbage. You should expect to fight infestations of face and cluster flies in early spring and house flies in late summer or fall. Fight them with a commercial flying insect aerosol space spray.
Spider: All spiders inject venom when they bite. Only two North American spiders are consistently dangerous to humans: the brown recluse, found in Kansas, Missouri, and the southwestern United States, and the black widow, common only to the warm southern tiers of the country. The recluse spider is identified by a violin-shaped mark on the top of its head. The venomous female black widow (the male isn't dangerous) has an hour glass design of red or yellow on the underside of its abdomen. Report spider bites

to your doctor immediately.

If you find the web of these spiders, don't attempt to knock it down. Spray the web with the residual insecticide first. Remove the webs after the spiders are dead. Any structures where the brown recluse or the black widow has been seen should be treated promptly with a residual oil spray containing lindane, chlordane, or malathion. Spray liberally in dark corners; with ordinary house spiders, dusting is preferred—windows, basements and crawl spaces.

Mosquitos: There are over 100 species in North America. After mating, the females of many species need a blood meal to nourish their eggs—it's the females that bite. After they bite you, they usually alight on a ceiling or an upper wall area to digest the blood meal. Quiet water, such as ponds, puddles, or water trapped in artificial containers or the crooks of house and garden plants, is where they lay their eggs.

Use screens with at least 18 meshes per inch. Use a residual spray on the ceilings and upper wall areas; use space sprays for heavy infestations. Also, kerosene or an emulsifiable insecticide can be used to kill larvae in rain gutters, puddles, and other areas of stagnant water. DO NOT TREAT ANY AREA WITH AN OUTLET TO A STREAM WITHOUT THE APPROVAL AND SUPERVISION OF LOCAL CONSERVATION AUTHORITIES. CALL YOUR COUNTY AGENT.

Ticks: Ticks carry diseases such as Rocky Mountain spotted fever, Texas cattle fever, relapsing fever, and a condition known as tick paralysis which disappears after the removal of the tick. Generally, the only tick to commonly set up house in your home is the brown dog tick, and it carries no diseases. A blood-filled adult tick drops from a dog and hides in the upholstery seam, the wool of a carpet, or a crack in a baseboard. It's in these places that the young ticks hatch and develop.

To remove a tick from the skin, hold a lit cigarette close to its protruding end, or by applying a heavy concentration of moist salt, kerosene, or alcohol. The tick will withdraw without loosing its head. This is very important to prevent infection. While you get rid of the brood of ticks in the home, have a veterinarian de-tick your affected pets. Spray with an insecticide containing malathion or diazinon wherever the pet sleeps and in breeding crevices.

NOTE: Government regulations could change and result in the removal from pesticides many of the active ingredients named above. Check with your dealer for newly introduced substitutes that meet current standards.

Power Tools:

Many of us own and use small power tools in our garages to make those common household repairs. Many accidents occur each year, usually to the eyes and hands, which result in permanent injury or disablement. Here are some safety precautions that should be adhered to:

1) Protect your eyes. All it takes is a small, sharp piece of shrapnel, a tiny bit of debris, to be flung into the eye to cause permanent damage. Always wear safety glasses while operating any power tool. The glasses should be shatterproof, not just any pair of eye protection. If you wear prescription glasses, there are goggles available that will prevent anything from hitting and breaking your glasses. If the tool blows sawdust in your face, wear a face mask.

2) A tool with a three-pronged, grounded plug should always be connected to a grounded, three-hole outlet; if an extension cord is used, use it with only a three-holed type of cord. Using an adapter properly is important too. Remember to use an adapter between the outlet and the extension cord; not between the tool and the cord. You are safe to use double insulated tools in two-hole outlets, because they have plastic housings and two-prong plugs.

3) Sometimes careless individuals operate a power tool while standing in water, damp garage or basement floor, even wet grass. They never forget it afterwards; the shock is devastating and often lethal. Check before you turn it on.

DISASTER, EMERGENCY & MEDICAL PROTECTION

Look down and see for yourself where you're standing. You may have accidentally spilled some water, or knocked over your refreshing glass of Gatorade, and not even know it.

4) Carrying a power tool by its cord, pulling the plug from the outlet by its cord, or hanging the tool by its cords are the swiftest ways to break the wires inside the tool. This causes the tool to turn off and on intermittently while it's plugged in, and could present a danger to the user.

5) For those of you who'll repair a power tool yourself before sending it in to someone else to fix, use extreme caution. Unless you want to get seriously shocked, be sure you're using only factory replacement parts and re-assemble the tool to the manufacturer's specifications.

Air Conditioning Emergencies:

When your air conditioner will not start, make sure it's plugged into the outlet. If it still doesn't work, plug in something you know works—then you'll know if the outlet is operative. If your unit is powered by gas, check to see that the pilot light is on. For units run by electricity, confirm that the fuse or circuit breakers are functioning.

Heating Emergencies:

If operated by electricity, and the heater fails to come on, determine if you have a power failure. Again, check the fuses or circuit breakers. And, of course, for gas-operated heaters, check the pilot light or move the thermostat to a higher setting; also check the switches, fuses or circuit breakers.

Physicians: Your Life Is In Their Hands

There was a day, not many years ago, when if you were too sick to go to the doctor's office, he would make a house call. Today, even your own doctor may tell you, "If you cannot come to my office, go to an emergency hospital."

Once there was a family doctor and he treated every family member. He delivered the baby, and throughout the years, looked after everyone in the family. Today in the medical profession, though, you find specialists.

Many doctors are known as family doctors or work in general medicine. When a problem requires specialized training, you're generally referred to a specialist in that field.

If you don't have a family doctor, or require the expertise of the specialist, how do you locate one?

Under the heading of physicians, most doctors are listed in the Yellow Pages by their specialty. Not all doctors will take new patients. The really busy ones want only present patients, or persons referred to them by another doctor or one of their patients.

You can also receive referrals from the American Medical Association (AMA) or even from a local hospital.

How do you choose the doctor that specializes in your particular problem? The following is a list which discloses the problem and the type of specialist who deals with it:

ALLERGIST .... Allergies.
ANESTHESIOLOGIST .... General and local anesthetics; your surgeon will normally appoint this doctor.
CARDIOLOGIST .... Heart and blood vessels.
DERMATOLOGIST .... Disease of the skin.
GYNECOLOGIST .... Female sexual and reproductive organs.
NEUROLOGIST .... Brain and nervous system.
OBSTETRICIAN .... Pregnancy and childbirth.
OPTHALMOLOGIST ..... Defects and diseases of the eye.
OTOLARYNGOLIST .... Ear and throat.

DISASTER, EMERGENCY & MEDICAL PROTECTION

PATHOLOGIST .... Tissue and structural changes caused by certain diseases; also, medical examiners in the event of death.
PEDIATRICIAN .... For infants, children and adolescents.
PROCTOLOGIST............... Rectum, anus and colon.
PSYCHIATRIST .... Emotional illnesses and mental disorders.
RADIOLOGIST .... X-rays and radium diagnosis and treatment disease.
SURGEON .... Surgical correction; there are specialists in all areas of the human body.
UROLOGIST .... Male urinary and genital tract and female urinary tract.

If you are new to an area, or just don't have a good doctor, ask your friends to refer you to their doctor who specializes in your problem. County medical societies also have a referral list. When you're referred by any group, however, they have a rotating list of doctors who want to be on the list, and no judgments about the physician's competency are made.

As with all professions, there are good and bad in the group. Before you trust your life to any physician, particularly if surgery is recommended, *obtain more than one opinion.*

Medications: How To Save Money & Your Life

Billions of dollars are spent each year to buy prescriptive and over-the-counter medications. When your doctor prescribes a certain medication for you, he often orders it by its brand name, such as Darvon.

All prescriptive medicines have a generic name. The generic name for Darvon is propoxyphene hydrochloride. It is made by many companies and cost less and has the same quality as the brand name counterpart.

About 25% of the most frequently prescribed drugs are made by more than one company and sold under various brand names.

The cost of a "pill" can vary up to 1,000%, depending upon which pharmacy you buy it from or which drug company manufactured it. Usually most brand name drugs cost double or more than their generic counterpart. You are paying for the name and their advertising costs. Next time your doctor prescribes medication for you, ask that the prescription be made in the *generic* name rather than the *brand* name.

In California, a pharmacist can substitute generic drugs for a brand name drug, unless the doctor indicates on the prescription, "Do not substitute." Often it'll cost you less to have the prescription filled in larger quantities rather than renewing it when you run out.

You almost always risk some danger or reaction when taking prescribed medication. Some can be very dangerous, unless the drug is used as prescribed and with caution.

Don't consume alcohol, or drive, or work certain machinery, or mix medications, if there is a warning on the label, or your doctor tells you not to. Use of illegal (or other) drugs can also endanger you.

Unless your doctor approves, don't take sleeping pills, tranquilizers, antihistamines, pain medication, anticoagulants or antibiotics together with the medication prescribed.

For your personal safety, *never* use someone else's medication. Not only is it illegal, but it could endanger your safety, even if the medication was to combat your symptoms.

Most of us store our medication in the medicine cabinet in the bathroom. That isn't wise. The reason is that the moisture in the air can result in the drugs' deterioration.

Keep all medicine out of the reach of children. Don't store drugs near a washer or dryer, as the heat can change the effectiveness of the medicine.

DISASTER, EMERGENCY & MEDICAL PROTECTION

Dispose all outdated or unwanted medicine in the toilet. If the medication is unlabeled, it may be to your best interest not to take it. Don't take medicine from its container while in the dark or when you are unable to read the label.

A final word: we know it's difficult to get your child to take medication. Many parents make the mistake of telling the child, "This is like candy." If you do, the child may want "more candy" someday and overdose.

KEYS INTO ATTACKER'S EYES

CIGARETTE LIGHTER INTO ATTACKER'S EYES

NEWSPAPER INTO ATTACKER'S ADAM'S APPLE

HEAL OF SHOE INTO ATTACKER'S EYE

KICK INTO ATTACKER'S GROIN

KNEE INTO ATTACKER'S GROIN

IV
EVERYDAY SELF-DEFENSE

Common Self-Defense Weapons

You are about to be attacked. Perhaps the attacker has only robbery in mind, but the attack may lead to rape or even murder.

When he planned his attack, he had no concern about what would happen to you. So when you defend yourself have no concern about what you'll have to do to him.

Throughout this book we highly recommend carrying a pocket or purse size tear gas weapon. That is, if your state has legalized it. Every state should, but a few have not. Bad Guy has access to any weapon he chooses; guns, knives, or whatever; but the law-abiding citizen has little or no self-defense weapons.

Once attacked, you have to decide whether to *submit* or *resist*. If you submit, you may be violated, physically hurt or killed. If you resist, you may escape. You have little choice; but, still, the decision is yours.

If you don't have a tear gas weapon, you may find a "weapon" that will fend off your attacker—something as common as your keys. Keys can be extremely effective. Take two or more keys and place them between your fingers. Make a fist and aim at your attacker's eyes! This sounds terrible, but remember, it's *your* life or his. If you miss the eyes, scratch his face with the keys. It's not as effective, but may put him off-guard and give you a chance to escape.

You smokers: when you're walking out at night alone, light a cigarette. When under attack, push it into the attacker's eyes. Better yet, if you can light your butane lighter, "flick your Bic" into his eyes or face. It may give you an edge and a way of escape. If you carry a book, hold it with both hands and shove the edge into his nose or throat. Or, a newspaper rolled up can be painful if thrust into the Adam's apple or stomach.

There is no limit to the contents inside a purse that can work to temporarily disable an assailant. An attack to the eyes with your rat tail comb; shooting hair spray into the eyes, hitting the eyes with your high heels, your umbrella, or anything in easy reach. Sharp kicks to the groin, of course, takes his attention off *you*.

Don't forget these: grab, twist and yank his hair; biting his fingers, neck, arms, ears, lip or chest; scratching your fingernails across his face, neck, arms or hands; if he wears jewelry around his neck, yank on it or twist it until it chokes him.

We have no intentions of making you into a martial arts expert. What we're going to describe and illustrate on the next few pages is very basic self-defense tactics. They're simple and easy to use by almost anyone if they are willing to spend just a few minutes to learn then practice them.

Why practice? You need to do them automatically. These movements need to substitute all of your helpless, panicky reactions that were learned while growing up. Practice in front of a mirror. Regardless of the defense-assault you use, remember your *body weapons:* head, hands, thumbs, fingers, feet and elbow. And where is your target? It's always the most sensitive part of the body which is most accessible to your defense—usually the eyes, nose, groin, and knees. But don't forget your voice; it, too, is a psychological weapon.

1. WHEN ATTACKED FROM BEHIND

2. SNAP HEAD FORWARD

3. THEN, SNAP IT INTO ATTACKER'S FACE

During an attack, yell "Fire!" or "Child Molester!" These "yells" are frequently recommended by self-defense experts. "Rape!" or "Mugger!" as we've said in other parts of this book, don't receive response as frequently from passers-by. But what fellow human could not answer to *this* cry for help: "HE'S KILLING MY BABY!"?

Attack From Behind

In an attack from behind, there are a few moves which help you to escape. If the attacker has his arms around you in a "bear hug," your arms restricted, without hesitation, *throw your head forward, snap your head into his nose and face.* The suddenness surprises him, and the pain he feels often makes him break his hold. Then, remember to *scream and run.*

Another method: the attacker has you in a "bear hug," your arms held to your sides. What you do is this—and it has to be done almost instantaneously to put him off: *fold your shoulders* in and forward, then immediately *back again,* and then you simply *drop down* from his hold and *throw your elbow* into his groin. This doesn't take strength. Leverage and surprise makes this work. Practice this move, carefully, with a friend or spouse—but skip the elbow in the groin part.

Here's another way to break an attack from the rear: as soon as you're in his clutches, *grab the assailant,* if you can, anywhere accessible with a free hand—this will give you support—then *kick back your foot, aiming for the kneecap,* then, as hard as you can and in one motion, *scrape your heel* from his knee down to his shin and *slam your foot* onto his toes. It's all one move. *Back into his knee, down the shin, and slam the toes.* Practice this, but do it slowly at first; get your coordination, because, remember, you aren't able to really see what you're doing all the time. Learn to do it without watching your foot. If the assailant still hasn't released you after your defense, throw back your head into his face, and repeat the above action, if necessary; in most cases it won't be, though.

Anytime an attacker puts a strangle hold on you (his arm around your throat), to avoid injury to your larynx or cutting off your breathing, merely *turn your head* so that your chin is in the crook of his arm.

Going for the genitals usually results in extreme pain for the attacker, but is an often misused defense-assault. Genital attack is not always recommended, but if you do proceed with this defense, here's how: if the attacker is behind you, *reach between your legs* with your strongest hand, and *grab the attacker's testicles, squeeze hard, then yank.*

One of our female instructors in the Nick Harris Detective Academy, who is a black belt, told us of another unorthodox defense that is the reverse of the above.

"The attacker expects his victim to yell," she said. "Or scream or faint. But if you place your hand, very gently, on his testicles and massage them, he's thrown off-guard, and you suddenly make your defense-attack, whatever way would be most effective." We can't say for sure that this works but in theory it sounds reasonable.

Attack From The Front

If you're attacked from the front, or around the legs bringing you to the ground, you're often able to simply reach out, *take the attacker's head* in your hands and *gouge the eyes with your thumbs,* pressing firmly. We know it's a difficult defense to use; most of us find it hard to reach out and actually touch an attacker, and gouging his eyes sends screaming shivers down most of our spines, but few defenses are as effective in temporarily disabling him. Tell yourself you're squeezing air from a beach ball. The pain is so severe, and the blinding so disabling, to save your life, you should use it—forget about your squeamishness.

Kicking or kneeing the *groin* has a time and place, and receiving an attack from the front is the time—it can be

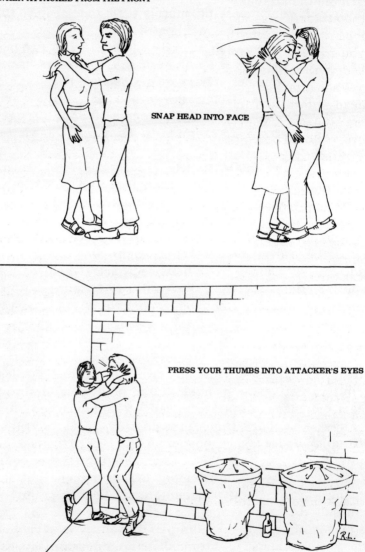

used conducively almost anywhere, but in tight spots like vehicles it can be very awkward. Two ways to successfully disable the attacker by striking the groin are: 1) Close to the attacker, *use the knee,* and 2) Two to three feet away, *kick with the foot.* Always repeat this assault-defense until you believe you can escape. His reaction in the face and body will tell you this.

When attacked from the front, we suggested you use your head—literally; from the front, you use your head to butt the attacker's. Your forehead is stronger than you think. In professional boxing, it's often used intentionally (although against the rules) to inflict additional injury to the opponent. *Anytime you use your head,* whether the attack be from the rear or from the front, *clinch your teeth and keep your mouth firmly shut.* If you don't, you may find yourself with a broken jaw or some missing teeth.

The *heel of your hand* is strong. You can easily execute a sharp blow with it. Direct it just under the bridge of the attacker's nose; it's very painful. At the same time, once contact is made, use *your fingers* and *dig* them into *the eyeballs.* A second choice for a target is *the chin.* In both cases, though, a heavy thrust with the hand to the nose or chin can render the assailant unconscious.

In a confined area, where there is limited room to kick or thrust out with the arm or hand, your *elbow* becomes the perfect weapon. It's strong; just open the palm of your hand and gently strike it with your elbow—it's like rock, isn't it? *With your elbow flexed,* you want to *jab backwards, aiming for* his *nose,* his *eyes,* anywhere in *the face:* even his *ears;* the side of his *head;* the *pit of his stomach* is another good elbow target. Regardless of the area you're in, the elbow attack is very useful and incapacitating.

The Voice

Experts tell us that the voice is a very dramatic weapon in itself. Anytime you're under attack and retaliate with an effective self-defense move—*yell*. You've heard the karate yell used by experts in demonstrations, and some of you probably thought it was for show. It isn't. Using all your voice's power, by taking a breath and forcing your diaphram in while you yell, you can create a startling reaction from even your biggest attacker. It's psychological, but it works. Not every time, but use it anyway. When you slam down your foot onto his toes, or kick him in the groin or knee, or throw an elbow into his face—it doesn't matter—use all your power by yelling; it'll give you extra energy in your defense, giving it extra power, force, and, above all, a feeling of self-confidence.

Self-defense against the rapist unofficially recommended by many police officers: if ever you find yourself in the midst of a rape attack, here's what you can do. Most likely, but not always, part of that attack is oral copulation. So if your attacker is alone, and you're forced to cooperate, provided that a gun or knife isn't being used in the force, *follow his request*. When he orders you to place his penis in your mouth, as many of them do, ask permission to *fondle his testicles;* if he complies, you *bite as hard as you can* on his penis, while at the same time you *rip off his testicles*. It's simple to do. The rapist will be in extreme pain, and your escape should be easy. Have no pity on him; he had none on you.

EVERYDAY SELF-DEFENSE

Tear Gas For Self-Protection

Are You Prepared?

Tear gas may someday save your life. We hope it will never be necessary for you to find out, but if that day comes, are you prepared?

If you answered yes, you may be fooling yourself. Hundreds of thousands of persons in California received the mandatory Department of Justice two hour training course on the use of tear gas. We conducted our own poll, and learned that most students had forgotten basic principles they were taught. If they encountered an assailant, it would be disastrous.

Instructional training isn't required in other states. What follows in this chapter is an exclusive and comprehensive lesson on the proper use of tear gas. We haven't found a more complete written work for the layman. For you who received formal training, this will refresh your memory, and provide additional, updated information. For everyone else, this will be your guide. Everything you need to know about tear gas.

We are certified instructors of the Department of Justice, and have trained thousands of men and women.

Before your lesson, we wish to take a not-so-imaginary encounter, one that you someday may really face. This is about *you*. At this very moment, would you be ready to face this encounter? Are you informed and prepared?

* * *

Late in the evening, you hurriedly walk alone in the direction of a multi-story public garage. It was as busy as Grand Central Station earlier in the day when you parked your car on the sixth floor. You weren't concerned then about the news item in yesterday's paper, THUGS ATTACK WOMAN WALKING TO CAR. But now you shiver thinking about it, because the assault took place in this very garage, and late in the evening.

Step-by-step you force yourself inside the darkened garage, deserted by the friendly parking attendants. Uncontrollably, your knees shake, as you ask yourself, "Why am I acting paranoid?" You aren't really alone, though, you have your bodyguard, so quickly you place your tear gas weapon in your hand, and approach the escalator.

Nervously, your thumb rotates the trigger from left to right, right to left. You can't remember which side is the safety, and which is the firing position. Halfway up the second level, you try to remember the range of the weapon, "Is it 10, 15 or 25 feet? Or is that the number of one-second shots?"

From above, you hear several muffled voices; automatically, your thumb rotates the trigger from side to side. You're prepared for the worst. "Do I fire into the eyes, or anywhere near the face?" You know there are two types of tear gas, the target isn't the same for each, and you're further puzzled by how to spray your weapon.

You feel a sigh of relief as the escalator moves slowly up toward your floor. Now you see the shadow of your vehicle parked in the distance. Suddenly, several young men wearing identical jackets and having identical intentions run at you; one in front of the pack has his hands held out ready to grab you, and in desperation you rotate the trigger to the left, press down and *fire*.

You made the *wrong* decision. You put your tear gas weapon into the safety position. You now belong to the thugs. You weren't prepared to face this attack.

* * *

If you know the basic principles of your tear gas weapon, you'll most likely disable attackers rather than become a victim. *Tear gas is safe, and very easy to use.* The weapon *is not* a toy, and must be treated as a very serious self-protection device. It really works: everyday, as private detectives, we bet our lives it does. Our weapons always go where we go.

The Controversy

Millions of persons were told tear gas may not work. It was July 2, 1981, the day *20/20* needed its eyes *examined*. Its "perfect vision" was as blind as an attacker would be when hit with tear gas. More persons watched this "report" that night than any other program on television. According to the Nielson ratings, it was number 3 of *all* television programs for the entire week.

They proved television can influence the lives of all of us. We'll never know how many thousands of viewers will be robbed, assaulted, raped or even murdered because they chose to throw away their only self-defense weapons, tear gas, or made the decision not to own one.

We must admit, the program was convincing, as telecast. They showed *brief* film clips of police officers "shot" with tear gas: only one went down immediately, the others during the few seconds shown on film stood up. Next, several ladies, during a mock-attack, "shot" police officers with no noticeable reaction. Convincing!

Now comes the real test. Let's view the *20/20* program in slow motion and stop motion. You see this all the time on TV sports, but not on *20/20*. This is how we viewed the playback: the police officers' eyes were closed after hit with tear gas. The officers "shot" during the mock-attack, again in slow motion and stop motion, were observed either jumping up out of the way of the tear gas, or immediately grabbing the hand of the ladies. Remember this: they knew in advance they would be shot with tear gas. In real life, this isn't the case.

Another important fact is that in all cases only seconds of the shootings were shown on television. Most persons actually shot with tear gas don't fall down immediately, and some never fall down because the tear gas isn't like being shot with a bullet. However, this non-lethal weapon disables the attacker with temporary blindness. Just a few seconds on film won't reflect this disability.

The nation's oldest detective training academy, Nick Harris, was asked by *20/20* to assist in the production. They falsely said that the program would be educational, and the final concept would depict that tear gas is a good self-protection weapon. We permitted them access to all of our training facilities. They spent several days with us, and remarked to the various instructors that the classes were excellent.

If tear gas doesn't always work, as reported on *20/20*, why then did the two producers of this segment, and the television crew, AT THEIR OWN EXPENSE, choose to be trained and licensed by Nick Harris Detectives? They *have* their tear gas to protect themselves. Perhaps the judgment on the effectiveness of tear gas was made in the editing room. We'll never know.

This is a fact! To prove the point that tear gas is extremely effective, Nick Harris Detectives conducted the most famous clinical study ever witnessed. In the presence of news media, two volunteers were instructed to injure co-author, Milo Speriglio, and the Academy's tear gas coordinator, Heather Lee. The media reported the test was authentic. One "attacker" was shot with CS, and the other CN, and within 3 seconds or less both were disabled. Millions of persons saw this on local and network television.

We allowed *20/20* to use this film. They showed a few seconds of it, and reported doubt of its authenticity. Hilly Rose is one of the nation's best known radio talk show hosts. After the *20/20* airing, he asked Mr. Speriglio to be a guest. Hilly believes in and carries tear gas. The program aired in Los Angeles on Gene Autry's station, KMPC. The discussion turned to Nick Harris Detectives' film used in their classroom. He pointed out that *20/20*, using this film, said tear gas doesn't work, yet just a few weeks ago, *ABC-Nightline*, which showed the same film, declared tear gas did work—*That's Incredible*. All three are on the ABC network.

The odds were more than 5,000 to 1 that one of the callers to this talk show would be Kevin Gibson. He is an athletic type, in his mid-twenties. He was also a student of the Nick

Harris Detective Academy the night *20/20* was filming. The California Department of Justice recommends that students (at their option) test the effectiveness of tear gas.

Usually only one person volunteers for this test, which is enough to demonstrate that tear gas really works. Kevin volunteered. We put just one drop on a cotton swab, placed it *under*— not in—his eye, for just one second. The effect, considered about 1/100th the affect of actually being tear gassed, was drastic. The angered young man told the radio audience, "I am 6'3", 185 lbs. Within seconds after Milo put a Q-tip of tear gas under my eye I was blinded, helpless and felt nauseous." The entire class witnessed this—so did *20/20's* producers, and film crew—they even filmed it. But not one second of this real-life demonstration on the effectiveness of tear gas was shown on television. As a matter of fact, nothing positive establishing the real value of tear gas for self-defense filmed by *20/20* at the Academy or our office was shown to the public. We were hoodwinked by *20/20,* but more important, the entire nation was, too.

The aftermath caused by this program can't be blamed solely on ABC-TV, its sponsors, or even the general management of *20/20*. They hire producers to independently write, produce, and film segments to be aired on the network. What takes place during the final edit can totally change the original story.

If you were persuaded by this *20/20* segment, that's your own choice, but we sincerely recommend that you accept the advice of real experts. Tear gas *does* deter crime. Criminals who have a choice between burglarizing one of two houses—one with a barking dog, the other without— will take the silent one. If two people walk down the street, a lady with tear gas exposed on her purse strap ... a man with tear gas on his waist, and two others without a display of tear gas—the criminal will choose the unarmed as his victim.

We believe tear gas will deter crime. One of the most noted journalists of modern time, Mike Wallace, on the CBS radio network, interviewed us on the subject of tear gas for self-protection. He told the nation that (reported) rapes in California has recently decreased. California is still the rape capital of the nation, but we strongly agree, rape has been detered because thousands of females in this state have armed themselves with tear gas, and the criminal has a healthy respect for this weapon.

Criminals acknowledge that tear gas is their enemy. When they see it, the weapon becomes a deterrent; when used against them, it becomes a force. Nearly all of the nation's tear gas experts totally disagree with the position shown on *20/20*. Today, tear gas is an important weapon used by law enforcement. To our knowledge, after the airing of *20/20,* police officers throughout the nation continue to use tear gas. Even the tear gas division of the California Department of Justice told us the public was misinformed by *20/20.*

Everything You Need To Know About Tear Gas

Q. Can tear gas kill?
A. No, it's non-lethal. There is no permanent damage. It doesn't paralyze anyone; it temporarily disables the attacker, and affords you the opportunity to escape. The attacker will be disabled for 5 to 45 minutes. The average is 20 minutes, but sometimes, without first aid, the effect will last over an hour.
Q. How difficult is it to shoot tear gas?
A. If you can operate hair spray, or an aerosol can, you can easily shoot your tear gas weapon. It's as simple as *press* and *spray;* it's safe to operate.
Q. What happens to the attacker when he's sprayed with tear gas?
A. It varies depending upon the type of tear gas, the body condition of the attacker, and other factors. Tear gas' major effect is temporary blindness. The eyelids force closed, and, if an attempt is made to open them,

the pain is more severe. During our clinical test, *20/20* did not show the volunteer in our film say, "It feels like thousands of needles going into my eyes."

When the attacker is hit, as directed, depending upon the weapon, we found subjects totally disabled within 1 to 3 seconds.

Normally, in addition to the blindness, the skin starts to burn, and there can be a stinging sensation. Often, there is nasal and saliva discharge, some coughing and gagging. Want more? It's not unusual to see the assailant panic, or feel he is suffocating, or dizzy, or have a headache and vomit.

Once hit with tear gas, some persons blindly charge. As soon as you fire at the attacker, move aside as if he is the *bull,* and you are the *matador.* But continue to spray him until you're satisfied he's disabled. You can use all of the tear gas in your canister, but it may not be necessary.

Q. How many shots does my weapon hold?

A. The smallest popular size—three inches long outside of the holster, CS tear gas—has an average of ten one-second shots or bursts. The next size has 15 shots, and is one inch longer. Most CN tear gas units are four inches in size and hold 25 shots. In many states, police-size weapons can be legally owned, in CN or CS, but for most of you, they're too big for your pocket or purse. They're six inches long with a diameter of the size of a half dollar. They have approximately 60 one second shots. It's an ideal weapon for home, vehicle or at work.

Q. What range will my tear gas travel?

A. *Do not* rely upon the range shown on your weapon's label. Most pocket or purse size (3 and 4 inch) weapons are labeled, "10-foot range." Some really go about 6 feet, 10 is nearly the average, and occasionally they exceed 15 feet. The police-size weapons claim up to 25 feet in almost every case. However, they fire at least

15 feet, and we found they often will fire around 25 feet.
- Q. How true are tear gas advertisements, or information on the labels?
- A. Often untrue. Weapons sold in California must be tested, and approved by the Department of Justice. Usually, their information shown on the label is correct. But, in other states, no test requirements are needed. Beware of false advertisements. One claims it has 50 shots — impossible for a three-inch weapon with a maximum capacity for 10 one-second shots. We see ads stating, "No aim is required." Totally false. Some weapons sold as "tear gas" really have no tear gas, just red pepper and oil; it'll stop a dog, but only make an attacker mad. Read the label. The minimum amount of CN or CS tear gas should be .09% to 1% in California; up to 2% in other states. Any of these weapons will disable an attacker.
- Q. Is all tear gas MACE?
- A. No, but both the public and press media often refer to any type of tear gas as MACE. It is a brand name of a product, perhaps the most popular.
- Q. Can an attacker protect himself from being tear gassed?
- A. Yes, but not under normal conditions. He can wear a gas mask or goggles, but he'll draw a lot of attention to himself. If he wears glasses, CN vapors will immediately attack him; with CS, particularly with the mineral oil base, the droplets fall from the forehead to the eyes, and cause blinding. Attackers with contact lenses have no protection against tear gas either. A stocking mask? Nope. Won't help.
- Q. Does tear gas penetrate through clothing?
- A. Yes, with rare exception. But, and this is most important, many persons are misinformed about this fact— YOU MUST fire your first shots at the face (see CN and CS methods below). Then, if you feel it's neces-

sary, you fire at the attacker's groin. Pain in other parts of the body will be minimal. Tear gas will react immediately when shot into the face and eye. It takes longer—seconds to minutes—to penetrate through clothing. We advise our students not to keep their tear gas weapon inside the pants pocket. One of our students told us his experience: while traveling on a long trip, he accidentally stuffed his weapon in his pants pocket. Other items pressed against the trigger and it discharged. He remembered the first aid discussed below. He removed his trousers, his underpants, then ran to the nearest gas station for the water hose. Police almost arrested him for indecent exposure.

We've been told tear gas (CS) with mineral oil will cling to the pubic hairs (as well as a person's beard), and cause pain far greater than being shot in the face. But remember, your first shot is at the face and eyes; the groin is your second shot.

Q. Where should I keep tear gas?
A. This is vital. If you're about to be attacked, unless your weapon is within *immediate reach,* it probably won't do you any good.

Ladies: your personal weapon must be close at hand, either clipped to your purse strap or placed *above* the contents of your purse.

Men: if your holster has a clip, attach it to your belt or trousers, or keep it in your shirt or jacket pocket—not in your pants pocket.

Q. When should I have my weapon in my hand?
A. This is very important. Place your tear gas weapon in your hand, and have it in the ready-to-fire position anytime that you are alone and/or in a potentially dangerous situation, in the streets, going to or from your house or vehicle, in an elevator, garage, sitting on a bus bench, in a subway, *anywhere* you're a target for any type of attack.

Q. Do I need more than one tear gas weapon?

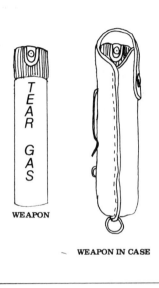

WEAPON

WEAPON IN CASE

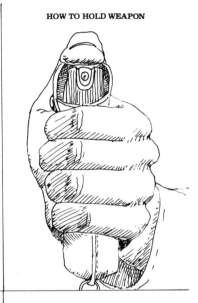

HOW TO HOLD WEAPON

TEST FIRE WEAPON

FIRING WEAPON AT ATTACKER

A. One will provide you with minimal protection. It's best to own several. No one, on a day-to-day basis, can take the weapon from room to room, and wherever they go. Tear gas should be treated like a fire extinguisher. Everyone in the household knows where it's kept, and it's never moved.

For your safety, one tear gas weapon should be permanently kept on the bedstand. The police type is best. By each door, hang a weapon by the holster key ring. Before you open the door to anyone, even persons claiming to be the police, a priest, delivery persons, anyone, reach for your tear gas, hold it in your hand until you're satisfied they're who they claim to be. Frequently, they're imposters—criminals who want to get in your home.

Q. Explain the controversy between the two types of tear gas weapons.

A. One is called *CN*, the other, *CS*. Their generic names are Alphachloroacetphenone and Orthochlorobenzalmalononitrile, respectively. But what is important to you is their true capabilities and limitations.

Both CN and CS tear gas are very effective. Many consider CN the easiest to use. CS is considered the longer lasting and stronger of the two. We believe, for maximum security, you should have both types.

The chemical used in most CN aerosol weapons will cause a vapor to be formed which will cover the attacker's face, when aimed at the face or just below. In most cases, this will quickly disable the attacker.

Your aim doesn't have to be as good with CN. All blind persons we trained and licensed were told to use CN. Some elderly persons or handicapped persons, who find it difficult to steady their hand and aim at a target, feel CN is best for them.

For most persons, we found CN may be the best tear gas for the nighttime when it's very dark and difficult to see your attacker. The effect of the tear gas will be less than that of CS. If the person receives first aid (see below), the effect will wear off in about five minutes.

CN is a water base spray. We don't recommend firing CN when it's raining, since the water base CN tear gas comes off with water.

CN vapors when fired into a strong wind, will often come back at you. While the attacker will receive the bulk of the tear gas, you may find it uncomfortable and receive some tears, but you still can escape.

CS tear gas' two major bases are acetone and light pharmaceutical mineral oil. The latter is our preference. To fire any CS weapon, aim for the bridge of the nose. One or more droplets must enter the open eye of the attacker for immediate effect. This can be accomplished by most persons without great skill. Many of the CS products marketed after July, 1981 have a better shotgun spray; firing it is as easy as CN. Even with poor aim, it's very difficult to miss the target.

We have found that persons shot with CS remain disabled longer, and the pain and blindness will be more severe than with CN. Most CS products have an ultra-violet, invisible dye that marks the attacker for later police identification. Often both CN and CS tear gas won't be effective on some persons under the influence of alcohol or narcotics, or some insane persons, as well as some animals. This isn't always the case; it depends upon many factors. If the person or animal doesn't show any immediate reaction to tear gas after a successful spray, they may not be affected. Many persons under the influence of the drug PCP can be hit with a baseball bat and not feel any pain, yet tear gas, on many occasions, has forced the eyes closed. While they couldn't feel a burning sensation, the eyes remained closed, and gave the victim an opportunity to escape.

Q. If tear gas isn't 100% effective, why carry it?

A. No self-protection weapon is perfect. A gun can jam, you can miss the assailant—anything can go wrong. An addict on angel dust (PCP) may not feel pain if shot with a bullet, and unless a vital organ is severed, or he bleeds to death, he'll still be a threat. Tear gas, however, is better than most anything you can legally own and carry with you. It's far better than no protection at all.

Q. What is the shelf life of tear gas?

A. In almost every case, two years. Not from the date you purchased it, but from the date it was manufactured. Most labels have a printed expiration date. We have seen some tear gas weapons claiming a life of five years, but it's best to replace any weapon after two years; the cost is minor. Tear gas is not refilled; the canister is replaced. The cost over two years is about one cent a day.

Q. Explain the safety and firing positions.

A. Most pocket and purse size weapons sold before July, 1981 had a safety. For those weapons, many manufacturers and experts recommend keeping them in the ready-to-fire position at all times, and not on safety. Experts also recommend that the tear gas weapon be kept in a holster or case, which could be fired without removing the top strap.

Almost all tear gas weapons sold with the case have what is called a top strap. It's a piece of leather type product, crossed over the top of the weapon where the trigger is, and locks with a snap button.

Once the weapon is inside the case, there is no reason to open it or remove the weapon to fire. Just press firmly on the top strap. For safety reasons, many manufacturers don't line up the weapon in the case. Assuming you are right-handed (reverse this if you are left-handed), hold the holster in your hand with your thumb touching the snap button. Aim in the direction of an imaginary target. Now, unsnap the button and carefully rotate the weapon inside the case so the firing pinhole (you can see it when the safety position is off) is in the direction of your target. You'll see the pinhole from which the tear gas will be discharged. Once it's aligned, re-snap the button. With your thumb you can rotate the trigger. In most cases, it's a white top activator. Similar to the type used on *Binaca,* the mouth spray. Rotate it all the way in the ready-to-fire position, the position in which your weapon is to be in at all times. Wnen exposed to children and you feel you must keep it on safety, rotate it as far as it'll go to the left. In this position it won't fire under normal conditions.

Hundreds of thousands of the pre-July, 1981 weapons had product malfunctions, most particularly the ones sold between December, 1980 through mid-July, 1981. A faulty valve was found to be the real cause. To correct this, one major CN manufacturer, *Curb 20,* eliminated the rotating safety and placed a top cap on the weapon. This cap prevents accidental firing, but also makes it more difficult to remove during an attack situation.

Some weapons, during a test fire, performed correctly, but within weeks failed to perform. Between July and August, 1981, many major manufacturers of CS introduced a more superior weapon. The safety was eliminated and the trigger was locked into the weapon, eliminating a major problem of it falling out of a unit having no safety case. They also produced a better spray, a real shotgun type. If returned, many of these manufacturers will replace, without cost, any defective weapon still under guarantee (one year is the average). In November, 1981 a more superior CS weapon was introduced by Aerosol Defense.

Most tear gas sold after January, 1981 was a clear or white gas. When test fired you'll see it most particularly in CS. This tear gas is the best on the market.

Tear gas sold prior to this date is often yellow in color. It's effective, but not as great as the later product. We advise discarding your yellow gas, and getting the white type.

Q. How should I fire my weapon?

A. Always press firmly on the trigger—or top strap if the weapon is in a case or holster. If you can, use your thumb. If arthritis or other physical problem prevents you from firing it, use your index finger, or if necessary, *both* thumbs. It's best to grip your weapon in your fist, wrapping your fingers around it. If you can, hold your arm out, direct it at the attacker(s), and spray, moving your hand side to side. You'll see the spray come out in a stream. We recommend you hold the trigger down until the assailant is disabled, or you run out of tear gas. Depending upon your weapon, fire as directed above. It may be best to fire just above your target, as most students, we found, aim too low. Then, move your hand in the direction of their face.

Q. Why should I test fire my weapon?

A. For your own safety. Test fire it as soon as possible, because we contacted hundreds of our past students who were instructed to immediately test fire their weapons, and we found many who didn't. Please, for your protection, test fire your tear gas before it's too late. At this very moment, thousands of you have tear gas weapons which will not work, because like any product, some are defective. From December, 1980 through May, 1981, tear gas manufacturers could not keep up with the nation's demand. During this period, more faulty weapons were produced than in anytime in this self-defense weapon's history.

Go outdoors when it's not windy, and pick any target—a tree, wall et cetera, staying clear of any person or animal. Aim your weapon as directed above. Press the trigger for one-half to one second, staying six six to eight feet from the target. If your weapon fails to fire at least six feet, and doesn't come out in a spray (shotgun pattern) but just a thin stream, your weapon should be replaced at once.

For various reasons, some of you will *not* test fire your weapon. We hope you'll never have to find out if your tear gas works or not. On July 20, 1981, retailers were notified that there may be a manufacturing problem with many of the following CS tear gas brands:

Chem-Guard
Life Guard
Protect-A-Life
Chemical Defense, and
Aerosol Defense

Check the brand name shown on the label. If your weapon is defective, return it to the retailer in exchange for a new, quality-control product. If they can't provide you with a new weapon, write to the manufacturer shown on the label.

The above brands were the only ones who notified us. Since most tear gas manufacturers use the same valve that was found faulty in thousands of units, they too may be defective. Most CN manufacturers use this type of valve.

Q. How important is a case or holster?
A. For weapons carried in the pocket or purse, and for your own safety, they should be kept in a case (holster). We prefer the type with a top strap, mentioned above. Holsters with clips are good, but small clips provided by many manufacturers, which attach to the canister, aren't recommended. We've found they can easily fall off. You can lose your weapon. Your holster should also have a rivet at the bottom for a key ring. We recommend that you carry your home and vehicle key on it; this will help you remember, *never leave home without your tear gas.*

Replace the key ring provided by the manufacturer with one that can separate your keys from the weapon. Remove the keys from your tear gas to open your house or car door, and hold the weapon in the other hand. Otherwise, when the key is in the lock, you can't immediately remove it and fire your tear gas.

Q. If I shoot an attacker with tear gas, what should I do next?

A. Leave the area at once, and notify the police when you're in a safe place.

Q. How old must I be to buy tear gas?

A. In most states you must be 18. Check with your local law enforcement agency. If a minor uses tear gas kept in a house, under the control of an adult, to stop an apparent assailant, by law, the minor may have broken the law, but in most cases they wouldn't prosecute the minor.

Q. Is it legal to carry tear gas on an airplane?

A. Federal law prohibits carrying tear gas on your person, in your check-in luggage, or in your carry-on luggage. This goes for both private and commercial airplanes. The law now reads: if you knowingly transport tear gas you shall be fined as much as $10,000. And if you willfully violate the law, the fine can go as high as $25,000 and, in addition, you could be sent to prison for a maximum of five years.

We don't know of any situations to date where the maximum fines or sentences were carried out. In most cases, when they're discovered, airport security confiscate the weapon. If a person carries tear gas aboard to hijack the plane, then the odds are he or she would receive the maximum penalty. We disagree in part with this Federal law. It states, "All tear gas materials are classified as hazardous materials and cannot be shipped on any passenger carrying aircraft." This is from the Department of Transportation Regulations, Part 175. 49 CFR. Tear gas is *not* a hazardous material.

We have reason to suspect that agents of the U.S. Department of Transportation and the Federal Aviation Administration, both in Los Angeles and Washington, D.C., agree with us. Thousands of tear gas weapons are shipped by air cargo everyday. We agree tear gas should not be carried into the airplane, or even in your carry-on luggage. But it should be legal to carry it inside your *checked-in* luggage as long as it is inside a double plastic baggie, or similar device. This law prohibits law-abiding citizens from bringing their personal protection with them when away from home. Today's tear gas weapons won't explode under high altitudes. The most they can do is possibly leak, and this is extremely rare, but if protected inside a baggie, there should not be cause for alarm.

Q. Where is it not safe to store tear gas?

A. Near a furnace or heater. Tear gas may leak if it has a faulty valve. Don't keep your weapon inside the glove compartment or trunk. First, you can't immediately reach it, and, second, it may be too hot. We have found tear gas usually can be kept inside your vehicle, even during hot summer months. Most tear gas weapons, as stated on most labels, are effective up to 120 degrees. Also, keep tear gas out of reach of small children, just like you would a gun, or medication. One other important tip: always fire your canister in an *upright position*. Shake it before firing, if possible. This will mix the tear gas with the carrier.

Q. When is it legal to fire my tear gas at a human?

A. The rule of thumb is: you can fire tear gas when you have good reason to believe that you or someone, regardless who, is in danger of great bodily harm. If you're alone in a dark area, and a stranger walks up and asks for the time, or a light, that alone is not cause to fire your weapon. If your purse is on a table in a restaurant, and a criminal picks it up and runs, that alone is not legal cause to chase and tear gas him;

however, if the purse is in your hands, and a purse-snatcher tries to grab it from you, and you fear he *can* hurt you, which often happens in this situation, then you can fire the tear gas.

You must spray the weapon at an assailant only while you, or others, are in immediate danger. If the suspect flees, you're no longer in danger, and by law, running after him now makes you the aggressor and him the victim. This doesn't seem fair, and we agree, but we didn't make the law, we're just reporting it.

This is the way we interpret the "rule-of-thumb" law: there are two threats you can defend yourself against. If, during these hypothetical threats, you, with good cause, believe great bodily harm would follow, then in the doctrine of the "reasonable man's thinking" (later described), you'd be justified in spraying the person(s) with tear gas in these examples:

VERBAL THREAT: "I'm gonna smack you up-side your head!"

ACTION THREAT: No verbal threat has to be included in the action threat. You can use your tear gas if a would-be assailant lunges at you, displays a weapon or a simulated weapon, one that you have good reason to believe is real. The weapon doesn't have to be in his hand, but in his immediate reach.

At the Nick Harris Detective Tear Gas Academy, we often use the following demonstration: the instructor places a 12-inch feather in his hand, walks among the students and says, "We will pretend we are in an alley, just you and me." He walks over to a student and says, "Against your will, I'm going to tickle you with my feather." As promised, he moves toward the person and tickles her. Then he asks the class, "Under the circumstances just described, and the action that followed, would this student, by law, be justified in spraying me with tear gas?"

The reactions vary; some say "no," some say "yes,"

and others are unsure. How would you have reacted? When the instructor said, *"Against your will, I am going to ..."* under California law he committed an ASSAULT, which is similar in most states. An assault is generally defined as an unlawful attempt (against your will), coupled with a present ability (the movement toward a person), to commit a violent injury on the person of another. During this demonstration a second law was broken—BATTERY. The two crimes committed together is called Assault and Battery, a phrase you've heard before. A battery is usually described as any willful and unlawful use of force or violence upon the person of another. How, you must wonder, can it be illegal to tell someone you are going to tickle them with a feather? This time, you think, the law is *out of order!*

Let's review the circumstances of this "assault and battery." As you recall, the two participants are in a dark alley. But it could be anywhere. You, the possible victim of a crime, are *fearful* because of the time, location, or light conditions of the area. Next, you are approached by a stranger. He holds in his hand what he *says* is a feather; it looks like one to you, but you can't be certain. He says he's going to tickle you with it. Was that all he really planned to do? He could stab you with its point, he could do anything he wants to you, with or without the feather.

A person was actually arrested and sent to jail in California for going around tickling persons with a feather. He was charged with Assault and Battery.

If under these circumstances, or any other type, you had *reason to believe* that you or someone you would protect, was in danger of great bodily harm, most likely you could have legally used your tear gas.

Branches of your neighbor's tree encroach upon your property line, and in the fall, leaves fly all around your yard. You tell yourself this has got to stop and, in a neighborly fashion, you call upon the resident of the adjacent property. You're determined this won't become a fight, and ask them to please trim the branches.

"You can go to hell," the neighbor says.

Now you're mad as hell, and place your tear gas weapon in your hand. "This'll fix that bastard," you say. You go back, aim, and spray him as he answers his door. What you just did is called *misuse* of tear gas.

You weren't under any physical threat; you weren't the victim in this case, your neighbor was. The legal issues concerning the use of misuse of tear gas is very important. This is what we tell our students:

"The courts have ruled, a person being assaulted is justified in using as much force as is necessary for self-defense. However, if the force is excessive, beyond what a reasonable person would have done under the same set of circumstances, then it's possible you can be charged with misuse of tear gas. If you shoot someone with tear gas without legal cause, you can be arrested, and/or sued."

Generally, you have cause to use any weapon, including tear gas, to offset the danger you face. It's important, however, that the amount of force used is both *reasonable* and *necessary*. For your personal protection, MAY THE FORCE BE WITH YOU.

Q. If I shoot an attacker with tear gas, must I render first aid?

A. No. But it's important that you know the first aid procedures. The label on your particular weapon should provide first aid instructions. These instructions may be required by law, or given by the manufacturer; however, they may not be 100% accurate.

Based on our knowledge, and actual clinical tests, we determined: with *CN,* flush the eyes with copious amounts of cool water, and do the same with other parts of the body affected.

The subject should be placed outside facing fresh air. A fan of any type, even handmade, can offer relief. The skin can be washed with soap and cool water. First aid is quite effective with CN. We found the major effects, like temporary blindness, will dissipate within five minutes following first aid treatment.

The first aid is the same when shot with *CS*, but *time,* not *first aid,* will erase most of the effects. First aid will give some comfort, but persons receiving mineral oil base CS gas will recover within an average of 20 to 45 minutes. Wash the hair thoroughly if hit with CS mineral oil; otherwise later a shower can cause tear gas still on the hair to drop to your eyes.

Whether shot with CN or CS, don't rub the face or eyes, for this will produce further bad effects. *Never use any kind of cream, salves, oils, or first aid ointments.* This can cause additional pain and blistering. Even sunlight, we've found, extends the temporary injury. In most cases, persons hit with tear gas will fully recover within an hour or less, and there should be no permanent injury.

Q. Is tear gas legal in every state?

A. Unfortunately, no. But it is in most. At this writing, California's law is the best. Anyone, 18 or older, not convicted of a felony, assault, or misuse of tear gas, can carry the weapon. A state license will be immediately issued upon completion of a two hour training course.

We strongly believe no one should have a tear gas weapon without proper instruction. That's why this chapter is so important. Studying this should give you the basic guidance on how, why, and when and where to use tear gas. The State of New York prohibits any citizen from owning or carrying tear gas. Manufacturers have told us, except for California, they sell more tear gas in New York than in any other! We even consulted the New York City Police Department and asked, "If a person licensed in California visited the Big Apple, and was confronted by one of your muggers, and this person used their tear gas in self-defense, would you arrest them?" The reply was, "Not if they're law-abiding citizens."

New York's law is clear: tear gas is illegal there; but not all police officers will follow the letter of the law when it favors the criminal, and works against the people. Believe it or not, New York State is among the unsafest for its inhabitants, and tear gas should clearly be permitted there. Minnesota also banned the use of tear gas until citizens became outraged, and forced a change in the law in April, 1981. In Nevada you can possess tear gas, but you must first obtain a permit and show "cause" why you need it. Bring a copy of the daily newspaper, reporting the last night's crime—this should be cause enough.

We devote an entire chapter on crime against tourists in Hawaii. In Honolulu, the capitol, tear gas, at this writing, is prohibited. Presently there is a move in the local legislation to permit persons to carry this self-defense weapon. It is, however, legal on the island of Kauai. We know of no other states which prohibit tear gas, but it's advisable to check with local law enforcement agencies in your area, or the ones in which you'll be traveling. Laws will change, hopefully, and all fifty states will "permit" citizens the right to protect themselves. Criminals don't fear the law. They carry any weapon they choose.

Our laws are not strong enough to stop criminals, maniacs, or even terrorists from robbing, assaulting, raping, or killing us. Our overworked, understaffed police officers can't catch most of the law breakers, and our crowded jails can't hold the small percentage of criminals caught, tried and convicted.

EVERYDAY SELF-DEFENSE

Your Only Protection Is Self-Protection

The most precious of all human rights is life. Yours, ours, and others. While you may not yet be aware, *we are at war.* We have a common enemy, not overseas, but right in our neighborhoods. This is an American Civil War against those who attempt to deprive us of our right to be safe at home and in the streets.

We have two choices; either *submit* and fall victim to the criminal, or *defend* ourselves against the enemy. Early in our tear gas training we coined this phrase: "Before the police arrive, your only protection is self-protection." Think about it.

What you do for yourself before the police arrive can make the difference between life and death. We know of no other media presentation in recent history that suggested to a mass audience to surrender and throw their self-defense weapons away, than did the *20/20* segment on ABC-TV. As we said, as far as we know, the producers and film crew have their tear gas weapons, they can protect themselves, but what about the hundreds of thousands of viewers they influenced? Don't permit *20/20* or any other unqualified source to sway you into laying down your arms, and perhaps the best means of self-protection. Tear gas, now that you know how to use it, will be your 24 hour bodyguard, and it won't get in the way, never needs feeding, and doesn't talk back.

Roone Arledge, President of ABC News, was provided an advance copy of this chapter, to rebut the 20/20 segment. No rebuttal was submitted.

The Pros And Cons Of Handguns

People are afraid. With cause. And they want to be protected. With crime waves frightening everyone, handgun sales across the nation have skyrocketed like never before.

The Second Amendment to the Bill of Rights makes clear that we have the (legal) right to bear arms.

A "Saturday Night Special" killed Senator Kennedy. Kids make zip guns. Anyone can buy a handgun from someone, anytime they wish. They can buy it immediately, or, in some states, following a five or more day waiting period, can purchase a weapon from a gun store.

There are many individuals and groups who want to ban all handguns. Their theory is good, but it can't work. If every handgun was taken away from *every* citizen, excluding the military and law enforcement, then many would endorse it. But that's impossible!

The criminal would still find ways to obtain guns, and the citizens would go unprotected. We don't encourage or even recommended that everyone own a handgun. In most states it's against the law to carry a concealed gun outside of your home or business.

We strongly believe in the constitutional right for citizens to own a gun, but discourage anyone from depending upon a gun for self-protection, UNLESS THEY THOROUGHLY KNOW HOW TO USE THE WEAPON.

We estimate at least 85% of all persons in the nation owning handguns never fired one. And many of them know little or nothing about the safe and proper use of the weapon. This could be disastrous to its user.

Many experts say that most persons unfamiliar with a handgun will hesitate before firing it at an assailant. We agree. Even if you do succeed in firing it, unless trained, you may miss hitting the assailant. Or, as in many instances where an automatic weapon is used, the person using it doesn't know that the first cartridge must be cocked into the chamber before the weapon can be fired.

By all means, own a handgun, but obtain training before attempting to use it to protect yourself. Improper instruction may cause you to have the weapon taken away by your assailant and possibly used on *you*. If you have youngsters of any age who will be exposed to your handgun, please keep it unloaded, instruct your children of the dangers, and keep guns out of reach. Never, ever allow anyone to "play" with an unloaded gun. Each year, hundreds of people die because they handled a gun they *thought* was empty.

For most persons, owning a gun is not the solution. A tear gas weapon, non-lethal, yet capable of immediate disablement, is something most persons can use for self-defense. If properly trained, its use can be easier than a handgun. You press a button, propelling a spray; you know it won't kill; it's much easier under stress than squeezing a trigger on a gun.

BOOK II
CONSUMER PROTECTION

I
SCHEMES & SCAMS

Interview With A Con Man

In the following interview with a professional con artist, specializing in money scams, we've left in details which explain how scams of this type are done successfully. Our intention is not imitation but specific information that may make you aware and informed of the processes that take place in money games such as short-changing businesses, ripping off insurance companies, or making small currency into larger bills. Information is your only defense against these quick, clever thieves. DO NOT ATTEMPT THESE METHODS. THEY ARE ILLEGAL.

John is 25 years old. He's from a suburban area, but moved to the city about high school age. It was then that he began learning his future trade.

Short-Changing

Q. How long have you been a con man?
A. I've always been a con, played money games. Like if somebody sells me something, and I want change for a fifty, and the item costs $9.00, I would give you two twenties, take my product and give you nine bucks

and get fifty back, change. So I paid one dollar and still get my fifty.
Q. Run that by us again.
A. Let's say you have a fifty dollar bill and I ask for change. And you have something for nine bucks that I want to buy. Now, I would put down two twenties, tell you to give me my product, and then ask you for change for the fifty and I give you two twenties. And you give me the fifty, plus the product.
Q. Do it again.
A. It's very easy. The best thing to do is to keep all your money. I had it tried on me. This guy came over to my house and wanted change. He had two twenties and a ten. He wanted a fifty. I said okay, and he put two twenties down on the table. I still had all of my money in my pocket before I even started to deal. I've dealt with this type of person before. So he wanted some stuff, and I gave him the stuff for $9.00. Which he didn't pay for. I gave him the stuff and asked him for the $9.00. I got my $9.00 and then I gave him the change. They try to confuse you.
Q. Go through the whole thing, and pretend I'm the store owner. Act like you want to buy this coffee cup for $9.00.
A. I would ask you for change for fifty during the process. That's the worst thing right there, because I'm taking your mind off everything.
Q. So you ask for the change at the time you lay the item down on the counter. So, I'm going to give you change for $50.
A. You're going to give me a fifty dollar bill, and I'm going to give you forty-nine bucks.
Q. Forty-nine?
A. Yeah, because the cup is nine.
Q. Okay, you give me forty-nine, and I give you fifty and the product. Don't I notice?
A. We've gone through this now three times and *you're*

still having trouble—so what do you think? Of course they don't notice.

Q. So, you just hand them forty-nine dollars, but you give them two things to think about. Buying the product and making change.

A. People can't chew gum and walk at the same time.

Q. And while you're doing that, you're also talking about something else, right?

A. Yeah. It's just quick thinking.

Q. And, if you make a mistake, they can't prove anything, right?

A. That's why you never give people money until everything is out. I keep all of my money away, until I get everything out. I'll never just whip it out. I do one thing at a time so I never get confused. I deal with other people's money, in and out, and it just ain't mine. If I get ripped off, I'm in trouble because I owe other people that money.

Q. Have you ever been ripped off?

A. Not personally. Once I was held up at somebody's house. I opened the door and there was a shotgun and a guy with a ski mask. I put my head in the chair. I saved a bunch of money for the people; it was a drug deal going down I guess, and he told me to stick my head in the chair, face first, so I grabbed the money off the table and sat down. I saw blond hair coming out of the ski mask; blue eyes and a shotgun. They took everything and left.

Q. They just walked up to the door and knocked?

A. Yeah, I answered the door, and that was it. It was all set up.

Q. Oh, they knew that the drug deal was going down?

A. I didn't even know about it. It scared the shit out of me.

Q. Where did the money eventually go? The money you grabbed off the table?

A. In my pocket.

Insurance Fraud

Q. Why don't you tell us about insurance jobs?

A. Okay. What we do is to take a BMW 320i that's legally owned and totally insured; theft, collision, etc. You report the car stolen. You have somebody steal it. For instance—you give me the key to your car, and you say you're going out to dinner at the Castaways. You go, and in between that time I come out, and when you come out, your car is gone. You have witnesses and everything. Right there your car is gone, so now you report it stolen. It's recovered by the police and back on the streets with identification on it. But it's stripped; seats and radio are gone, back windows are out, rims and tires are off. Fenders to trunk, everything is gone, and it's not worth much totalled. So the insurance company gives you the first chance to buy it back if you want to. Or they give you full value for the car. If your car is worth $6,000, they give you their $4,000, plus your car. Since you had it done, you have all your parts back. You put your car back together, which is legal, and if anybody asks, say you bought the parts from a junkyard. So you have your $4,000 from the insurance company and your car, which has been put back together with your own parts. And you insure it again. Not with the same company because you'd get cancelled for reporting something like that. It's commonly done. It's done with houses too.

Q. You take the items away, but you don't destroy them. They disappear, you collect the insurance money and put the items back, and no one knows.

A. The only one hurt is the insurance company.

Q. And the premiums go up.

A. So? I'm not insured.

False Identification

Q. We hear you can make an I.D. for people—how?
A. You just go down to the Hall of Records, find somebody who died, get their birth certificate and go to the Department of Motor Vehicles. You say you're that person and get your picture on the license.
Q. You just show them a copy of the birth certificate and they accept that, right?
A. The Hall of Records' death and birth records aren't kept in the same place. They're kept in two different places, so they don't know what's going on. You find somebody who was born the year you were born, or whatever year you want to be born. If you're 18, you'd want to be twenty-one. Get somewhere close to that date, someone who died when they were a baby so there's no social security number on them.
Q. You want to get somebody who died as a baby?
A. Young, a young age so there is no social security. Because you can get caught from social security numbers. Even if they were 12 or 15 years old, they still might have had a social security card.
Q. When you take it down to the Department of Motor Vehicles, they don't ask to see the original?
A. No. They just want to see a copy. You take your test, get your picture on the name, and you've got yourself a license. You go get your social security number, you get a credit card, and charge it on his name. Charge everything. Give it to your girlfriend. Yeah, credit cards. I personally used to know a mailman who would give me credit cards before the people even got them, and we would charge up, and they would get a bill. They didn't even know they had a card.
Q. How would something like that work? Would you split it with him?
A. Yeah, I'd buy him something. A pair of skis, whatever. I'd just go and charge it.

The Money Game

Q. Are there any other common con games?

A. Sure. The money game.

Q. How does it work?

A. It works best with a white person and a black person. First, the white guy goes into a liquor store and buys some beer or something with a twenty-dollar bill. Now the bill has a phone number written on it ahead of time.

Q. Then what?

A. After a few minutes, the black guy goes in and buys something with a one dollar bill, something that he'd get change back on, like a pack of cigarettes, and then he leaves. But only a few seconds go by, see, and he's back up at the counter. The person behind it remembers the black guy, you know, because he was just there.

Q. Then he says what?

A. He says, "I got my girlfriend's number on that twenty I gave you. And you only gave me change for a dollar." The guy behind the counter might argue with you, but he looks in the register, and sure enough, there it is, right on top—with the telephone number just like he said. He isn't about to doubt it. And so the guy gets nineteen bucks back.

Q. Is it that simple?

A. Yeah. Like taking corners from twenties and putting them on ones. Tagging it's called.

Q. What about the clerk who just throws a bill into the register and then gives you change?

A. They are taught to take it and put it on the cash register so there won't be a question about it later. If I see an inexperienced person drop my ten right into the register and give me change, I would immediately say, "I gave you a twenty." I jump on him. It's already in the register, so he doesn't know for sure. They have no proof, that's why they're taught to keep the money on top until the change is given.

SCHEMES & SCAMS

"Making" Money

Q. Tell us how you change a one dollar bill into a twenty.
A. Oh, "Making" Money?
Q. Tell us about that. We believe it's called counterfeiting.
A. What we do—what I do—is get a good crisp one dollar bill. You've seen a few of these that were passed right off on you, and you didn't even know it. I cut the corners off of my twenties, just like this. (He demonstrates.) Then I glue them onto the ones; you cut and shape them perfectly. Then when you play your money, it's all a money game, it looks really good. You just get that sucker right on there. But you need a good one dollar bill. You gotta hold your Georges. You've got to pass it at quick liquor stores, in and out places, or gas stations. You hold it like this (his thumb over George Washington). I have some at home that I've already made up, but I don't do too much. I know people working at gas stations and they know what I'm doing. So, I don't really pass too much phony money. For one dollar, you get $19 change.

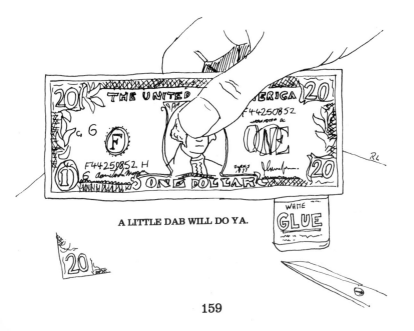

Q. You use four different twenties? You won't take them all off of one bill?
A. No, it's a waste. You can only take a couple of corners off for it to still look good. I take only one but you wouldn't take it without the four corners. So, now I need two more twenties to make me a "legal" bill.
Q. Can you only do it with twenties?
A. You can go fifties or one hundreds, but you need fives or tens to do that. But "Washingtons" are very common. I never do this to pay a debt. I do it at cash registers. When I give them the twenty, I show them I have other twenties.
Q. When you give them the phony twenty, don't they spot the George Washington?
A. No, they just give you change, and then they bury it.
Q. Do you try to distract them while this is going on?
A. Sure. You keep talking.
Q. You can't feel it when you touch it?
A. You can feel it, it's thicker.
Q. Do you use glue?
A. Yeah, Elmer's white glue. Just a little dab, and then flatten it. Just a little dab will do ya. (He laughs.)

Fraud: The Con In Confidence

A loved one of yours has died. The funeral is published in the obituary column of the local newspaper. Perhaps a week has passed and your grief has lessened, but the very thought of that loved one's passing affects you.

While reading one morning at the breakfast table, the doorbell rings.

"Good afternoon, may I speak with such-and-such," a middle-aged man wearing a suit says to you, smiling. You stand silent for a moment.

"He passed away last week," you tell him. "May I help you?"

"I'm really sorry to hear that. But I just came to collect the $138.50 on the encyclopedias he ordered last month."

Not wanting the loved one leaving the world owing a debt, you pay it.

You just fell for a major con. The con man got you by taking advantage of a weak period in your life. During periods of sorrow don't let them catch you off guard.

Another popular con: you receive a phone call from the "bank examiner." He tells you, "We're investigating a dishonest bank teller and we need to catch him in the act." He tells you that he needs your cooperation. "We marked all of the money in the teller's drawer." The bogus bank examiner meets you in your home. He tells you to make a large withdrawal, and to turn the money over to an inspector for examination with an ultra-violet light. At the bank, you make the withdrawal and turn the money over to the "inspector." You're ordered to remain in the bank lobby, while the inspector checks out the currency. You wait and wait and wait, but no inspector returns. When you approach the manager of the bank, you learn there's no real inspector, bank examiner, or suspected bank teller. You were used. And what about your bait money to catch the thief? Well, you just gave it to the real thief, so say goodbye to it.

There are hundreds of con games that separate you from your money. Don't get caught by "The Sting." Be alert to anything that sounds too good to be true.

Even though the motion picture, *The Sting,* opened many people's eyes, even today, hundreds of thousands of people fall victim to the con artist every year.

Here is another one. A man approaches you. He looks dishonest, turning his head around, looking to make sure no one watches.

"I need cash quick," he tells you. He rolls up his sleeve, and displays a few men's and women's expensive-looking watches on his arm. You take a quick look at the brand names. You think you see a Bulova and other famous names. You know they must cost over $150.

He shows you a few "diamond" rings on his finger. Worth hundreds, maybe thousands you tell yourself. You love a bargain, every one does. But you look at the shady salesman, and you know deep down, it's "hot" merchandise, but you can't resist the "steal" of a lifetime.

So you dig deep into your wallet, give the man only $40.00 for the "Bulova" watch, and another $60.00 for the "half carat diamond ring."

He leaves and you look around to make sure no one is watching. You stuff the merchandise into your pocket. Then you're home, carefully examining your treasures. Upon close examination you find that it's not a Bulova, but a cheap imitation, spelling the name with a slight deviation. What about the ring? Well, what you bought was a very cheap imitation of a "diamond" mounted on a brass ring without any gold.

The street salesman did nothing illegal. He even carries a resale permit to sell merchandise on the street. He didn't tell you it was a Bulova, or that the ring was made of diamonds or gold. All he said was, "I need cash, quick," and you gave it to him. The watch, purchased wholesale with a band, cost him about $4.00. It will work for perhaps a few months, if that. You paid him ten times its value. You paid $60 for the ring; his cost was about $3.00.

So if someone offers to sell an item far below its actual value, be wary. By gaining your confidence, the rip-off artist can tempt you with all sorts of bargains.

Ripped Off Renters

A client retained us to locate a man who had ripped him off of hundreds of dollars.

SCHEMES & SCAMS

Here is his story:

"I was searching through the classifieds under 'Homes for Rent or Lease.' I spotted a 'bargain.' I read it twice to make sure it wasn't a printer's error. It said, CHOICE ENCINO PROPERTY, three + two, den/pool; $485. mo., 555-1748.

"I knew that similar homes in that area were renting for $900 or more, and I could not figure out how it could be so cheap. Was it a fixer-upper? I dialed the number, and I talked to a man who told me that he wasn't as concerned about the amount of rent as he was the tenant. He said he wanted someone financially stable, someone who was responsible.

"I thought, 'there really is a Santa Claus.' So I went and met him and he said I was the ideal tenant and that I fit all of his requirements. He gave me the address, and I went and saw it, and the house was more than I ever imagined.

"He spent a few minutes qualifying me, before he handed me a 'standard lease.' I read it, I signed it, I paid him for the first and last month's payments, plus $200 for the security deposit and cleaning fee.

"He told me that it would be vacant in 30 days. I thought the wait was worth it. I didn't know that waiting was another check in his favor. And I have never seen the money since and I'll never see myself as a tenant in that house."

If you'd been this man, this is how you would have gotten ripped off:

When you signed the lease with Con Man Joe, or whatever name he is using at that time, and wrote your check, in good faith, you knew the property wouldn't be yours for 30 days. During that 30 days, Con Man Joe cashes your check.

You never think that the "owner" is a con man. And neither did the fifty or more persons who responded to his classified ad. All toll, people gave him about $88,000. When the 30 days came to an end, the moving vans lined the streets, everybody ready to move into the same house.

How did he do it? The con man legitimately leased the house from the real owner, and paid $2,100 for the first and last payment, and security/cleaning deposit. With the $88,000 he collected from the ripped-off renters, his net profit was about $85,900, less $60 advertising costs.

You should be suspicious. Sure, the bargain is too good to turn down.

So before you part with any amount of money, be certain you're giving it to the rightful owner of the property.

If the person renting/leasing the property says he/she is the owner, insist on examining the Grant Deed. Moreover, talk to the neighbors, ask if "Con Man Joe or Jane" is really the owner.

You can also verify ownership by checking public records at the county tax assessor's office, or calling a land title company.

If you are offered a "sub-lease," which is to lease from

the person who holds the master lease, be sure that you have the approval of the rightful property owner to sublease. Today, rentals are not bargains. Perhaps you just might find one, but that will be rare. The cons are *very* common.

Energy Scams

Through a friend of yours, a firm contacts you and makes an offer to invest in the marketing of a "miracle car." Their guarantee of high returns for your dollar, and your curiosity, have you inviting their representative over one evening for a talk in your home. The representative tells you that this miracle car will deliver a fuel system capable of over 100 miles per gallon.

"It will make the VW look like a gas hog," he tells you.

You ask to see some indication that this car does exist. He then produces a very professional-looking, detailed drawing of the miracle car in blueprint. There is even an actual full-scale model in his display of pictures.

Then he convinces you. He says, "Not everyone can participate in this venture."

By the next day you have withdrawn all of your savings and have "invested" it with the miracle car manufacturer. He leaves you some fancy stock certificates and his number and tells you he will contact you in two to three weeks.

A few weeks fly by and you're already planning your retirement. Your impatience forces you to call the miracle car company to find out when you'll be getting your first dividend check. After a couple of rings you hear a recording: "The number you have dialed is no longer in service ...." You try again, and again, but you know you have dialed it correctly. Your future financial plans are shattered.

It was all a sham.

* * *

The 1980's is the era of energy conservation. As a result of this, some persons make fortunes, while others lose everything they own.

Today the con man and phony promoters have lured unwary investors with false promises of high return profits and special tax breaks.

With the ever-rising cost of oil and gas, owning "a piece of the rock" of an oil well sounds like a smart investment.

Another major scam uncovered by the Securities Exchange Commission was a fraudulent selling of oil and gas leases to investors with a "promise" that within two years they'd recover the initial investment, followed by the profits "gushing up."

The series of oil wells *should* produce an average of ten barrels of oil a day, the investors were told. However, the promoters knew in advance that wells previously drilled by others in the same field were averaging less than one barrel per day. Not worth the time and cost of exploration.

The truth of the scam is this: the investor—perhaps you—or someone you know— would have to wait at least *eight years* to get back the original investment. In today's money market, the same investment would have doubled in a Certificate of Deposit account.

Please don't think that your level of naiveté would never allow this to happen to you, because every year thousands of people have it happen to them.

Next time you run to the bank to make a withdrawal to firm an investment, walk slowly, and think.

Mail Order Robbery

If you pick up any newspaper or magazine and search through the classified advertisements, you'll find fishy schemes baiting you to cast away from a few of your dollars.

Some advertisements, of course, are legitimate. Some are outright fraud. But they border so close on being legal that many people just don't recognize a ruse when they see one.

Like this one:

> EARN $35,000 IN 45 DAYS—SECRET COSTS ONLY $10. GUARANTEED SUCCESS. Mail ten dollars to Get Rich Fast, Inc. P.O. Box ...

You might say, it's worth the gamble. But is it? Thousands respond to such advertisements, and what do they get? *Instructions* .

That's right, instructions. Just like the advertisement said, you'll earn $35,000 in 45 days. And here's how:

"Place advertisements in your local newspapers, magazines, et cetera. Copy the ad that you responded to. When 3,500 people respond to your ad, you will have collected $35,000. And just like we promised, you can do it in only 45 days."

Another sampling from another newspaper in another town:

> OFFER OF A LIFETIME
> UNCUT EMERALD CRYSTALS
> $25 A POUND

Send check to Gem World International, Limited. P. O. Box....

Within a few weeks, your treasure arrives. Just as advertised, the emeralds are green. They're natural, uncut: 50 uncut dusty chips and pebbles—ideal for the bottom of your fish tank—and their value? About fifty cents.

Maybe you can be fooled by gems, maybe not, but most of us know nothing, or next to it, about gems' value.

Now gold is another story. Everyone knows its value per ounce because it's reported in the papers daily: $400 per ounce, $600 per ounce, $850 per ounce. We're told by every kind of media, so we *know* what gold is worth today. Or do we?

YOU STRUCK IT RICH! YOUR PERSONAL GOLD MINE ONLY $99 FOR A GENUINE 24K ONE GRAM GOLD INGOT! Send your check today. Limit 5 per person. Gold Mine, U.S.A., P.O. Box ...

Why only five? Why not fifty? Or a hundred? Well, by placing a limit on what you can purchase, they know many people will think it's a bargain and buy all they can. If you have ever responded to this advertisement, you'll remember how angry you were at *yourself* for being ripped off. You could've kicked yourself for not making a couple of phone calls, or checking out a book or two from the library.

You see, in this gold scam, one *gram* is equivalent to 1.543 *grains*; it takes 24 grains to make 20 *pennyweight*, which is one *ounce*. Sound confusing? Well, it is—even if you use a calculator.

The bottom line is this: you would have paid $3,000 per ounce of gold—far more than the current gold market.

This type of rip-off is not limited to mail order, either, because many similar ingots are sold in "legitimate" jewelry stores. Sometime in the year 2,000 this might be a real bargain, but for the 1980's only a fool would give five one dollar bills for one hundred pennies.

An encyclopedia on this subject still couldn't cover the mail order rip-offs. Our point is, don't expect something for nothing, and don't think a small investment will bring you riches. The only ones getting rich are the promoters, who lure readers into sending their money so that they can "get rich, too."

Remember the Pyramid Scheme: a few people got $100,000 for a $1,000 investment; but many thousands lost their $1,000, and got absolutely nothing for their trouble.

When you gamble in Las Vegas, the "house" has the odds

in their favor, but you have some chance of winning.

When you send your money to the mail order robber, your odds of striking it rich is a million to one—and, then, only if Lady Luck is with you! And a dozen four leaf clovers and a gross of rabbits' feet won't help you, either. We tried them.

Beware of Free or Cheap Government Land

"Mail $3.00 and we'll send you your 'Insider's Tips' on how *you* can get government land—absolutely free!"

Every year, thousands of persons answer such advertisements. Some end up paying hundreds of dollars for these "tips."

The fact is the government no longer gives its land away free. The government doesn't even sell it cheap. In most cases, if the truth be told, the land sold by Uncle Sam generally costs more than privately owned land nearby.

On an average year, the various government agencies sell less than 1,000 acres and these acres are sold in parcels through public auctions. So what insiders' tips are needed to make a bid? None that we know of. About the best inside tip you could have is what is inside your pocketbook.

There was a time you could homestead government property. That ended in 1976. The original Homestead Act of 1862 was repealed in every state but Alaska, yet today, it's impossible to homestead even in that vast, last frontier due to legal complications.

You won't purchase land for a song, so save your money by not responding to the promoter's claims. The more clever of them even use company names that give the impression they are part of the U.S. Government such as "The American Bureau of Land Distribution." Impressive. But bunk.

Fix-It Fraud

There are a lot of con games that separate victims from their money. One major con is called the "roofing fix." The target: often an elderly widow.

Your door bell rings: "Hello, I'm Mr. Good Guy, the roofing expert. I just finished a job on one of your neighbors' roofs, and noticed yours needs some minor repairs. It's best to do it now, before the rain, and since I'm already in the area, you'll save a lot of money."

Nice guy you think. He's so concerned about you. "Let me take a closer look," he suggests. After a few minutes, he climbs down the ladder.

"Better come on top of the roof. I'm afraid it's much worse than it looked from a distance."

Mr. Good Guy knows that few elderly women will risk climbing up onto the roof. You take his word. "Better do what you can, but I can't afford to spend a lot." That was the reply he expected.

You hear bang-bang up on top of your roof, and see him busy saving your roof, and saving you money. Soon he finishes, and presents you with a bill.

"You really lucked out, only $379.40 for everything." Usually the woman pays.

In most cases, the roof was fine, maybe in need of a few minor repairs that an honest roofer would have charged $50 for. If you don't pay, you may be threatened with legal action. Some states, such as California, have a Mechanic's Lien Law. If you authorize repair on your home, and refuse to pay, once a court awards a judgment against you, a Mechanic's Lien is filed with the county recorder. Until the bill is paid, plus interest, and other cost, you can't sell your house. If it's left in your estate, your heirs must first pay off the lien.

Next time a repair person, of any type, who just happened to be in the neighborhood, wants to save you money, tell him thanks, but no thanks. If you need something fixed, you'll call a repairman.

The Great Pretenders

The "sting" master has a thousand bags of tricks. He constantly devises new ones. Beware of any offer that gets you rich fast—something for nothing, or nearly nothing—an offer you cannot refuse, or anything that just doesn't sound legitimate.

Almost all of us hide our valuables. For a few, the secret hiding place is under the bed, or under the sink; regardless of where, no one, except your heirs or loved ones, should be told where you keep your jewelry, cash, or other valuables.

Here's a con still being used today:

You respond to a knock on the door. Two men, clean cut, in suit and tie, await you. One flashes a badge: "Police Officers, Ma'am, we're with burglary detail. There's been a rash of crimes in the area, just checking." Almost looks and sounds like Joe Friday, from *Dragnet*. You know from the news media that crime is up in your area. You let them in.

"Don't mean to alarm you, but you may be one of the victims. Better make sure nothing valuable is missing."

You run to your hiding place. But wait, only special people should know where you keep your valuables, right? Well, the police, they're okay.

You look over the valuables, everything is intact; you feel relieved because the burglar missed you. The police officers suggest that as long as they're there, they should do a security home check for you. You agree, of course.

They look around and point out a few security tips. One officer takes you outside to look over the windows and doors, the other remains inside, continuing his security check.

After a cup of coffee they leave. You feel more secure now than ever. Perhaps your hiding place was not as good as you thought it was, so you decide to find a new location. Under the bed, you find your box of valuables, and pick it up. It feels empty. You open it, and everything is gone. And so are "Sargeant Joe Friday" and his partner. Those convincing "police officers" were con men. And you became their victim. Turn the clock back a moment.

Knock, knock. This time you tell the "police" you don't keep anything of value in the house. To verify if they really are the police, keep them outside and call the police department to see if there's a police investigation in progress. If the police aren't certain, they'll dispatch a patrol car to check it out. Remember, anyone can buy a badge, or even a uniform. Many legitimate police uniforms are stolen from dry cleaners.

Law enforcement bureaus in many cities offer a free home security check of your premises and give you advice on how to prevent burglaries. Most of the information given is contained in Book III *How to Protect Your Property*.

George Sunderland is the Senior Coordinator of Crime Prevention of the American Association of Retired Persons. For a free copy of "How to Spot a Con Artist," write him at 1900 K Street, N.W. Washington, D. C. 20049.

II
PAPER, PLASTIC AND POVERTY

Cash to Ashes

The last place you want to keep large sums of cash is at home. Many people do; either they don't trust banks, or they're careless. What happens in the event of a fire? Your cash is reduced to ashes. Do you kiss it goodbye?

Well, the U.S. Treasury Department usually can identify the ashes, and replace your money loss due to a fire. In Las Vegas, the MGM Grand Hotel fire reduced an estimated one million dollars in cash to ashes, but much of it was salvaged.

The Treasury has 30 highly-trained currency experts who properly identify mutilated or burned currency.

One man hid $500 inside the barrel of his shotgun. He forgot that he did, and went hunting, pulled the trigger, and the money blew into a thousand pieces. He retrieved the scraps and recovered most of the money.

The government checks over 50,000 claims each year and redeems over ten million dollars of destroyed money. If your money is burned or otherwise mutilated, first ask your bank if they can redeem it. If not, consult the U.S. Treasury Department, Bureau of Government Financial Operations,

DCS-BEPA: Room 132, Treasury Annex, #1, Washington, D.C. 20226.

If you owned a bar of gold, and cut it almost in half, one piece containing 49%, the other 51%, its value would be in proportion to its size. Not so with currency. Money, if cut or divided as shown above, would be worthless to the holder of the 49% and would retain its full value for the holder of the 51%. If you own *over* 50% of a torn U.S. currency note, it will be redeemed at full value by the government.

For over a quarter of a million debtors each year, there's another "burning" in the making ....

Warning Signs of Bankruptcy

Each year over 350,000 Americans declare bankruptcy. During your lifetime, five percent of your generation will take this drastic measure.

We interviewed experts who gave us the following guidelines to determine if you're a likely candidate for joining the ranks of the bankrupt.

Do you borrow to pay for the food on your table?

Have you applied for a new loan before repaying an existing loan?

Do you withdraw from your savings account to pay your everyday living costs?

Do you sell your family belongings to pay bills?

Do you write checks you know will bounce?

Do you take salary advances or cash advances from credit cards to pay your ordinary household bills?

Have you taken a second or third mortgage on your home, knowing you can barely pay on the first?

Do you plan to consolidate your debts to repay your other loans?

Can you barely get by with two incomes in the family?

Yes or no?

If you answered yes to any of the foregoing questions, you'd better start thinking about planning a budget.

There are many books on the subject. *Reader's Digest's How To Live On Your Income,* is highly recommended.

One thing you can be sure about: many families go bankrupt because they don't know how to use—or they abuse—their credit cards.

How To Save Money With Your Credit Card

Visa and Master Card are among the most popular credit cards in use today. They compute your interest charges on the average daily balance. If you pay your charge bill when due, there is no interest or late penalties.

The Federal Fair Credit Billing Act protects you. If you buy a product using the credit card, and later find it was defective, you can refuse to pay for it, until it has been satisfactorily repaired or replaced.

There is a restriction, however: the cost of the item must have been more than $50 and the purchase made in the same state as your residence, but not more than 100 miles from your home.

Billing is done by computers; like humans, they make errors. Often you're billed for someone else's charges, and sometimes the error is in addition. If you discover what you believe is an error, you have no more than 60 days after the bill was mailed to you (not when received) to protest the erroneous charge. The credit card company has 30 days by law to acknowledge your complaint, and another 90 days to resolve it. Your best protection is to send your complaint by *certified* or *registered mail.*

If they can establish that it is your charge by proof of your signature on the voucher, or other method, then it's likely you'll be required to pay the charges. But if they fail to acknowledge or resolve your complaint, you may legally not pay up to $50 of the disputed amount, whether an error was actually made or not.

If the credit card firm doesn't resolve the complaint according to the Act, you can sue for the actual damages, plus twice that amount (double damages) and the amount of finance charges imposed. If the amount involved is $100 and less than $1,000, you are also entitled to the cost of the lawsuit and reasonable attorney's fees if you're successful. Small Claims Court in each jurisdiction have various limits. In California it's $750.

You have 14 days after your credit card bill is sent before you're required to pay it without interest or penalty.

You're unlikely to see this, but hotels, restaurants and other merchants who take credit cards may, if they wish, give a discount of up to 5% if you pay by cash. They can't, however, add an extra charge if you use the credit card. The average merchant pays the credit card firm around 3% of the total amount of purchase, and in many cases, this charge is fixed into the cost of your purchase.

The days of receiving a credit card without requesting one are gone. By law, unless you make a verbal or written request for one, you won't be issued the "plastic card."

Responsibility for lost or stolen credit cards is covered in Book III, *How To Protect Your Property*.

Someday, a *single* credit card will replace cash, checks and conventional credit cards. Bankers have already forseen this trend. Use the "credit card" instead of writing a check, and the amount is automatically deducted from the deposit you have on hand with your bank. Remember, your piece of plastic is cash—your cash. Guard it like you would those ten, twenty, fifty and hundred dollar bills. Use it wisely. Don't buy more than you can afford to pay for. At 18% or more per year, you're paying almost one fifth more for the item purchased if you don't repay the credit card firm when billed, and choose, instead, to defer payment.

III
CONTRACTS & WARRANTIES

Warranties: Read Before You Buy

You're about to buy a product. You want to spend your money—hard-earned no doubt—for this product because you like it. The man on T.V. said it's great. The salesman tells you, "Don't worry, either, folks, it's all guaranteed!" And that makes you feel easier about buying it.

Guaranteed to do what? Fall apart in a week? Today there are warranties and there are warranties. But unless you understand the mumbo-jumbo about what's covered and to what extent, you can't "comparison shop" and get the best warranty.

Effective January 1, 1977, warranties on consumer products costing more than $15 must be available for you to read *before* you buy. So don't wait until you purchase the product, take it home, unbox it, only to find a guarantee which is "limited." Some products require you to mail the damaged item by registered mail along with $5 to cover the "costs" for handling and such other nonsense.

The new Warranty Act states that all warranties must be written in ordinary language, with no fine print, and no legal jargon. Every term and condition of the warranty has to be spelled out for you.

Beware of salespeople who give free advice about guarantees, because when it gets right down to it, you're only covered by what the manufacturer has put into printed words. Now, the Warranty Act doesn't require any manufacturer to give a warranty. In many cases you buy "as is" and pay the cost of repairs yourself. Can you recall the last time you read every word of a guarantee that came with a product you purchased? Most people don't read them at all.

What is the difference between guarantees and warranties? None. They mean the same thing: a promise by a manufacturer or a seller to stand behind their product. Of course, the kinds of promises you get can be as different as hot and cold.

What types of warranties are there?

A *Full Warranty* means a defective product will be fixed or replaced free—and within a reasonable time. You won't have to do anything unreasonable to get warranty service, like shipping a piano or a full-sized television set to the factory.

Your Full Warranty is good for anyone who owns the product during the warranty period. If the product can't be fixed, or hasn't been fixed after a reasonable number of tries, you get your choice: a new product or your money back.

Sounds good? Yes, you get a lot of protection, but the word "FULL" doesn't promise to—and doesn't have to—cover the whole product. It may cover only part of the product, like the picture tube of your T.V. set. Or it might leave out some parts, like the tires on your car. If your Full Warranty covers only the picture tube, and the sound goes bad, you pay. Always check what parts of the product are covered. Often, the parts that cost the most to replace or are known by the manufacturer to go bad early, are not covered.

Today, though, you still won't find many products offering the consumer a Full Warranty. Most products

CONTRACTS & WARRANTIES

which offer the consumer a warranty, provide for a *Limited Warranty.*

"Limited" in this case, we believe, means "be careful because something's missing." Examples: It covers parts—but not labor. The part may cost only a few cents, but the labor can be expensive. Sometimes you get a smaller refund or credit, depending upon the length of time you've owned the product. Like tires for your car. It may be a 5 year or 60 thousand mile tire, but if it's damaged in one year you may only receive 40% credit.

The Limited Warranties sometimes require that you return a heavy product (a baby grand piano) to the factory for service. And they can charge you for "handling," a ridiculous but legal fee they can charge to help offset their own cost. Limited Warranties can only cover the original owner, too.

Some products carry more than one written warranty. It may have a Full Warranty on part of the product, and a Limited Warranty on the rest. Take time to read the guarantee—every word of it—before you buy.

Next comes the *Implied Warranties,* which are rights created by state law, not by the manufacturer or company. All states have them.

The most common Implied Warranty is the "Warranty of Merchantability." This means that the seller promises that the product you buy is fit for the ordinary uses of the product. A reclining chair must recline; a toaster must toast; and a camera must take pictures. If they don't, you have a legal right to get your money back no matter what the warranty says.

Another legal warranty is the Warranty of Fitness for a Particular Purpose. If you buy a product relying on the seller's advice that it can be used for a special purpose, the seller's advice may create a warranty—but you have to prove it. Example: the salesperson tells you his sleeping bag will be suitable for zero degrees, and it surely isn't; you have a warranty against the seller no matter what the printed warranty says.

Implied Warranties come automatically with every sale, even though they aren't written out. If a seller says nothing about warranties, you get the Implied Warranties. A seller can usually get out of these warranties, though, by merely stating in writing that the seller gives no warranty at all. But a seller can't give a written warranty and get out of the Implied Warranties. If you receive a written warranty, you get the Implied Warranties, too, which may give you protection that the written warranty doesn't.

Service contracts may wrongly be called warranties. The warranty comes with the product at no additional cost. The so-called "extended warranty" is a service contract, and it *will* cost you more for the additional protection. Compare the free warranty with the additional cost service contract carefully. Look for how much more the service contract really adds.

Other warranties are spoken promises and advertising claims. You have the legal right, according to the federal government, to get what the company promises. Normally, these rights include the right to "consequential damages," which means the company must not only fix the defective product, but also pay any damage the product did. If your freezer breaks down and the food in it spoils, the company must pay for the food you lost.

Now the catch: the manufacturer can slip out of this by simply stating in the warranty that it doesn't cover consequential damages, in either a Full or Limited Warranty.

Read your warranty. Other times, the consequential damages cost more than the product itself. If your car's anti-freeze is faulty, and the engine block cracks, you may only get the cost of the anti-freeze—a few dollars—rather than the hundreds of dollars to repair the car. Think twice. If the consequential damages are excluded, how much damage can the product cause if it doesn't work? Compare warranties before you buy.

CONTRACTS & WARRANTIES

In recent years, product liability damage claims have cost manufacturers millions of dollars. If you are physically injured by a defective product, regardless of the warranty, you may have a right to get money for the damages, so consult an attorney.

When does the warranty end? You have a six month parts and labor warranty. On the seventh month the product doesn't work. Are you covered? Yes—but only if you complained about the product not working during the warranty period will the company have to take care of the problems, no matter how long it takes.

Read warranties *before* you buy, but only if you want protection and the best deal for your money. It may be worth paying more for a product with a better warranty. The extra money you pay can protect you from an even bigger repair bill. Always retain your sales-slip and warranty.

The Federal Trade Commission enforces the Warranty Act. Report any violations of the law to the Warranties section, Washington, D.C., 20580. This same commission enforces laws covering contracts, too ...

Contracts: A Binding Legal Agreement

What is a contract? It's when two or more parties agree to an issue, either orally or written. No special form of legal language is required, and all parties can be bound to the contract. Not all contracts are legal, some can be broken.

A legal contract is an *offer* by one party, and an *acceptance* from the other party(s).

Parties to a contract must be capable of entering into an agreement, so they can't be insane, under the age of 18, unable to understand the terms, intoxicated, senile or sign the contract under undue influence, duress or malice. *Consideration* is usually the key word. It's what you and the other party will exchange, in money, service or otherwise. Contracts for gambling debts are not legal. No act of any type involving illegal activities are enforcible.

There is no such thing as a "standard contract." Most contracts are written by an attorney to protect the interest of the parties he represents, not you.

Before you sign any contract, read it (including the fine print), and be sure you understand it. If you don't agree to every clause, ask the other party to strike it out. If you don't, you will be bound to the contract.

A noted physician retained Nick Harris Detectives to break a contract between he and his partner. The client had signed the contract without reading it first. When the Nick Harris investigator gave the client a Client Retainer Agreement to sign, he signed it, again, without reading it. The investigator just shook his head, and told the doctor, "If you'd read everything before you signed it, you wouldn't be here today." He read the agreement completely, and then signed it.

If a contract involves a lot of money, it would be well worth the cost of retaining an attorney to review it before signing. But, if you do sign something, you *do* have some legal protection....

The Cooling Off Law

People often act impulsively. That's why, in California, when you go to a gun store to buy a handgun, you have to wait fifteen days to receive it. You have to wait out the cooling off period. This law prevents many potential deaths, whereby, during a fit of rage, you're prevented from buying a gun. In many cases the person *did* cool off.

CONTRACTS & WARRANTIES

There now is another law protecting the consumer with the so-called cooling off rule.

* * *

Supper will be ready in fifteen minutes. Then your doorbell rings. It's Mr. So-and-So, the friendly salesman.

"Good evening," he tells you. "I need just three minutes of your time."

You explain that you're just about to sit down for supper.

"No problem, I'll just be a moment, I want to give you a gift."

He looks around your living room and finds some empty space, points to it, and says, "What I have for you will go perfect right here." He shows you a color photo of your "gift": a handsome "empty" bookcase.

Somehow he no longer appears to you as a salesman, but as a guest. Within a few minutes, you're convinced that you need a $900 set of encyclopedias, and sign a contract to buy them. After all, you tell yourself, the bookcase was free.

The salesman leaves, shaking your hand, and patting your dog on the head. You sit down to dinner and wonder, "How did he get me to sign that contract?"

Your kids have grown up, you hardly have time to read the newspaper, let alone an encyclopedia: $900 down the drain.

No, there's a way out.

A Federal Trade Commission rule permits the consumer to change his mind, *for any reason,* and may cancel the contract within three business days.

The purchase must total at least $25, and have been made in *your home*. Other exceptions include anywhere other than the seller's normal place of business, such as parties at private homes, conventions, etc. The rule doesn't apply to mail or phone sales, or for real estate purchases, securities, insurance policies or emergency home repairs.

By law, the salesperson must inform you of your right of

cancellation, and must give you two copies of a form to cancel. If you decide to cancel, send one copy to the business firm, and keep a copy for your own records. It may be best to mail it certified.

The seller, after receipt of your cancellation, has ten days to return your money. You don't have to agree to ship the merchandise back, but if you do, you accept the risk and expense. You can give the seller twenty days from date of cancellation to pick up the merchandise. If he doesn't, you can dispose of the property. However, if you don't make it available, or fail to send them back, and if you agreed to, the seller can hold you to the contract.

You won't have any problems if you simply remember this: before you sign a contract, cool off.

IV
PRODUCTS AND SERVICES

Shopping Myths

Paying For The Label: Brand vs Off-Brand

Most of us believe a Rolls Royce is better than a Ford, because it costs more. In this case, yes, but it's not always true. Many products that cost more are no better or even of less quality than a lower priced item.

"You get what you pay for" is not always the rule. Today's designer jeans are the in-thing. If you're buying for quality and not status, then you don't have to pay for the manufacturer's label.

Besides just paying more for a product, many persons buy well-known brand names rather than buying what may look like an identical item under a name not so well-known, or totally unknown, but costing much less. Many national brand items are made under one or more names. In most cases, they're manufactured identically. The brand name goes on one product, the lesser name goes on the other, and the difference? You pay fifteen to 50% more. The off-brand may be your best buy.

Discounts and Sales

True or false? You get just as good service from a full retail store as you would a discount store. True, discount houses may not spend as much time as you would like demonstrating the product, but if you know what you are buying from seeing it elsewhere, you'll save some money buying it from the discount dealer. If the product is backed by a manufacturer's warranty, it doesn't matter where you buy it.

What about an item bought solely because it's on sale or drastically reduced? You probably don't want it, at least at that moment. When the product is sold so cheap, nearly everyone will buy it, because they can't resist a bargain. A major chain store recently went out of business, and during its final three days, advertised 60% off list price on everything in stock. Thousands of people waited in line for the doors to open.

The bargain of a lifetime began. There was pushing and shoving, and everyone filled the store like a bunch of vultures fighting over carrion. Almost everything worth anything was already sold, yet this "Army" of your neighbors and mine cleaned out the store within minutes, buying everything in reach, mostly junk.

The price was right, but did they really need it? Stores don't give things away; seldom is a sale really *below their cost*. And beware: "Lost our lease" only means that they're moving or going out of business; they didn't lose their lease, *it expired*. Rather then spend thousands of dollars for advertising and paying a staff of salespersons, a store truly liquidating can sell directly to a liquidator, and avoid these extra costs.

Going-out-of-business-sales: beware. Particularly in the clothing industry. A case in point: one well-known men's clothing store, of which your authors were customers over the years, went out-of-business. It appeared to be a legitimate sale.

"Where's George?" we asked the salesman. He was the

owner of the establishment, and we knew all of his regular salespersons, but he was nowhere around, nor was his regular staff. "George who?" was the response. Something was fishy.

After a little investigation, we discovered our friend George had indeed quit business, selling his established firm to what we'll call "the bandits."

Here's how the scheme works. George has about $250,000 of actual inventory, and four months left on his lease. The "bandits" pay him $117,000, just below wholesale, for the inventory, the goodwill, the name and the store lease. They run big newspaper advertisements: "George Quits—Everything Must Go Regardless of Cost." His old time customers, of course, flock to the store for the bargains.

We noticed that only a fraction of George's regular merchandise was on hand, while the other items were things he never carried in the store. The salespersons looked like auctioneers, and didn't know a regular from a long; but then a sign clearly pointed out, "No alterations. All sales final."

"Bandits" mix the regular stock with pure junk, outlet type garments made in Hong Kong or elsewhere—seconds, irregulars, or last year's fashions, all over-priced. A $75 marked garment, "reduced 50%," looks like a real bargain if you thought George once sold it at that price; but he didn't. You "steal" the garment for only $37.50, and take home $20 worth of second-rate basement garments. Was it worth the bargain? Until you go home, give it a second look, or try it on, you won't really know. But then it's too late; all sales are final; and George isn't around to care anymore.

But if you really care about what you're buying—in this case, clothing—you should know about this ....

Buying Clothes

Ten-years ago you could buy a tailor-made suit for the cost of what an off-the-rack pair of trousers costs today. The average consumer spends about 8% of his yearly income to buy clothes. Beware of the recommendations of the salesperson. Many are untrained and inexperienced to help you make your decision.

Well-known brands often cost more. It's very popular today to wear clothes bearing a famous trade name or trademark. But are you paying too much? Perhaps yes.

Here is some good and bad about material used in clothes:

COTTON: It's strong, comfortable, absorbent. Unless treated, it will shrink. It wrinkles in the wash, unless it has a wrinkle resistant finish.

WOOL: It wears very well, springs back to its original shape, is warm, can shrink, and attracts moths. Some people are allergic to it.

RAYON: This material is weaker than most fibers. It's soft and comfortable, but may shrink or stretch, unless it has been treated. Light can affect its color.

ACETATE: This is weaker than most fibers. It has a silk-like look. Some colors fade.

NYLON: This material is strong and elastic. It can fade in sunlight. Nylon can be uncomfortable or clammy in warm or humid weather.

POLYESTER: One of the most common forms of material today. It's strong, and will resist most wrinkles, as well as stretching or shrinking. Like nylon, it can be uncomfortable in warm or humid weather.

SUEDE: So-called ultra-suede is excellent for jackets. This is perhaps the most wrinkle-free garment. Regular suede, or suede-type trousers, wear out very fast, and show wear, particularly in the knee area.

ACRYLIC: This material is weak when wet. It does, however, resemble wool in many respects. Acrylic will accumulate static electricity.

PRODUCTS AND SERVICES

Sometimes, even the most expensive and well-known brands are poorly finished. Much of all clothing is imported from countries where labor is cheaper.

Check each garment carefully. See if the seams are smooth and properly finished. Make sure the garment has room for necessary alterations.

Check the stitching. It should be even and low. The hem should be even and properly finished. Buttonholes should be neat, smooth and strong, and sewn through both sides of the material. The linings must be firm and smooth fitting. Zippers should be smooth, flat, and securely stitched.

Here is a little about buying shoes: if the shoe fits ... well, you know the story. Today, it's not unusual to pay up to or more than $100 for a pair of adult shoes.

Some bargain shoes can damage your feet. Look for shoes with neat finishings, because you don't want to buy a pair with rough edges, loose threads or heels, paper thin soles or exposed nails or tacks.

A good shoe will have reinforcement at points of stress, and will be made of firm yet pliable leather or fabric.

It's unwise to buy shoes from a mail order catalogue. Never buy by size alone. Often, different styles have different fits. Try the shoes on first. Everyone has one foot longer than the other. Make sure the shoe fits the larger foot comfortably.

The ball joint of your foot and the arch base of the shoe should meet at the same point. There should be one-half inch between your big toe and the shoe's inside tip. And be sure there is plenty of support for your instep.

As a general rule, a good pair of shoes won't have to be broken in. They'll fit you comfortably when trying them on in the store.

When you buy shoes for your children, they should always be tried on first. Your child's foot has not yet fully developed, and putting a misfitting shoe, or wearing a tight shoe, can cause irreparable damage to the foot.

Many wise parents buy children's shoes a half size larger

than the children's feet; they fit well and can grow into them.

The Five Grades Of Meat

Unless you're a vegetarian, this is very important to those of you who enjoy beef, lamb or veal.

There are five grades given by the United States Department of Agriculture:

PRIME: This grade is the most tender, juicy and, of course, the most expensive. It's rarely found in the average market. A gourmet butcher shop, very rare today, may carry *Prime*. This grade is sold to fine restaurants, because the meat has a high fat content.

CHOICE: This grade is juicy and popular with consumers, and quite tender and flavorful. It has a high content of fat, but is not equal to *Prime*.

With the recent change of grading laws, what was once called *Good*, today is called *Choice*, resulting in more meat with less fat and generally tougher. Yet, it costs more now.

STANDARD: This grade is really less flavorful than the above grades. It should only be cooked using moist heat, such as a pot roasting or basting process.

COMMERCIAL: The animal for commercial purposes is usually older. It must be cooked a longer time with moist heat, but the fat makes this grade flavorful.

More than one-half of the meat available in supermarkets is *Choice* grade. *Standard* grade can be more nutritionally beneficial to you because of the lower fat content.

Some markets sell ungraded meat, which is less expensive and is equally nutritious as graded meats, but less tender.

Bon appetit.

Mirror, Mirror On The Wall

"Mirror, mirror on the wall, who's the fairest of them all?"

Many women agree, that using cosmetics is important to them. But how much do you really know about the cosmetics you use?

There is a certain risk involved in the use of cosmetics. More than 22,000 persons are treated each year in the emergency rooms of hospitals for serious injuries. Others were treated by private physicians, and many went unreported.

Thallium acetate, well-known as a rat poison agent, was an active ingredient in a hair remover cream. It caused blindness.

Consumers now spend over ten billion dollars a year for cosmetics. Anyone can manufacture a cosmetic without pre-market clearance of cosmetic ingredients from the Food and Drug and Cosmetics Act.

Effective November 30, 1977, all cosmetics shipped by interstate commerce must, by law, list their ingredients. As a consumer, you can compare the ingredients with the more expensive cosmetic with an inexpensive one, and decide if you're paying more for the content or simply the brand name, packing and advertising costs.

Even if the label states "for sensitive skin," you must read the label and determine if *you* are allergic to its ingredients.

We'll tell you how to test the product; the manufacturer won't. To test the product and determine if you're allergic to the cosmetic, apply a limited amount inside your forearm. If you have a bad reaction, don't use the product. If the reaction is serious, see your doctor.

For hair dyes and permanents, a patch test is probably best. Put a very small amount of the dye on your forearm or behind your ear.

Don't scratch the eyeball when removing any eye cosmetic. Don't use any perfume, colognes or aftershave lotions if you applied a suntan lotion. This can cause splotching and irritation. Almost all feminine sprays are propellants. By holding the spray closely to the body, you can cause an irritation.

It's not advisable to allow children to spend an unusually long time in a bubble bath. Rashes and urinal tract infections can occur. And, of course, like medicine and weapons, keep all cosmetics away from small children.

Solid Gold

You look at the magnificent gold ring perched in its black velour case, and decide that it would look wonderful on your loved one's finger. As you admire it the jeweler tells you, "This is an investment. It's solid gold."

Beware: no jewelry of any kind, advertised or referred to as solid gold, is "solid gold."

To be pure "solid" gold, the metal must be 24-carat gold. If the jewelry was solid, it would be too soft and would bend or break. No jewelry is made out of solid gold.

Gold used in jewelry, at its finest, is 18-carat. This actually means it is composed of 75% pure gold, and 25% alloy.

When you see an advertised piece of jewelry stating it's "solid," under the terms of the Federal Trade Commission rules, this only means that the piece is not hollow in its center, and doesn't imply that it's 100% gold.

If a watch or piece of jewelry is listed as gold, it must be 10-carat or greater (10 Kt. or 10 Kt. gold). If it's marked 12-carat filled (or 12 K.G.F.), this means the plating is at least 3/1,000ths inch thick.

Next comes the 10-carat Rolled Gold Plate (or 10 K.R.G.P.), which is not less than *half* the thickness of gold filled in gold weight.

The word "gold" sounds expensive and valuable, but all

PRODUCTS AND SERVICES

types of gold don't have the same intrinsic worth. Not all "gold" is gold. If the phrase "gold tone" is used to describe a piece of jewelry, don't believe that it's made of gold; it means it's any yellow tone metal, or is thinner than the thinnest described gold in the Federal Trade Commission Guidelines.

So. Now you know how solid solid gold is.

How Safe Is Your Safe?

Today, many business people and homeowners own a safe. The contents often include: valuable papers, cash, checks, stock certificates, contracts, licenses, insurance papers (including fire and burglary insurance), jewelry, stamps and coins, and other important or valuable items.

There are two reasons for having a safe: to prevent destruction by fire, and to prevent theft.

Are we risking our valuables and treasures?

Perhaps yes. The loss of fire is greatly reduced, but not completely guaranteed, and the probability of a semi-professional or professional burglar getting his hands on your valuables while inside your safe is too real for belief.

First, let's examine fire risks. Most safes have a *fire tolerance factor*. This means it will withstand a certain degree of heat before the fire destroys the contents. Check the manufacturer's label to determine its heat resistance capacity. Inquire if its sufficient for most home/office fires. Your local fire department should be able to advise you on this matter.

You can further "insure" your safe's fire resistance capabilities by placing your most valuable items inside an inexpensive, fire resistent security chest. It's lined with at least one-fourth inch thick asbestos insulated material covering all surfaces. If by chance the fire penetrates through your safe, the contents are further protected within the security chest. Small chests cost under $35.

So far, so good. If you're threatened by fire, your safe may save your valuables, but what about the burglar?

Having any safe is better than no safe at all. Now comes the bad news. If the safe you purchase is small in size and weight, something anyone can lift, say under one hundred pounds, kiss it good-bye. The burglar won't even wait to crack it open, he'll just take the safe and all.

And if the burglar brought along a dolly (the type used for moving refrigerators, etc.) or "borrowed" the one you used to wheel out the garbage cans, then even a three hundred pound safe will be no problem for him to steal. And he doesn't even have to take the safe home. In most cases, he can break through usually within seconds or minutes. Rather than encourage would-be burglars, we'll not detail the methods but just touch upon common entries made by the typical burglar.

PRODUCTS AND SERVICES

The combination safes costing as little as $75 and upwards to $500 or more can be opened with limited skill. The locks can easily be broken where the thief, with a little patience or skill, can "hear" the clicks, and break the combination code. They have several means of entry.

Experts tell us that safes installed permanently in the ground of a home or office usually are the best. First of all, they cannot easily be removed, and, with clever preparation, can easily be hidden. This does not mean if it's found that the combination can't be cracked. Today's sophisticated burglars use metal detectors to find your "hidden" safe. Keep the location of your safe extremely confidential. Limit the location, and especially the combination, to only those with a *need to know:* your spouse, heirs, business partners, or the like. Don't tell friends, employees, or any others where you've hidden your safe.

If you can't afford the cost of a good safe, but want to protect your important papers and valuables, purchase a fire resistant security chest and hide it somewhere where it's least likely for a burglar to find. Remember, most burglars only have minutes to rip you off. They'll look in the obvious places first.

So, your three-foot wide by four-foot high safe in the den would be the burglar's first prize. He'll know this is where you keep your valuables, because you believe your safe is safe. Don't advertise your personal Fort Knox. Hide it.

Even though it may be a bit inconvenient, perhaps the safest safe is a bank safety deposit box. They offer no insurance against loss, but they provide real security.

Life Insurance: Investment Or Protection

Almost everyone has some type of life insurance. Few understand what they have, though. The policy is a guarantee. If you die, your beneficiaries (heirs) are guaranteed a certain sum of money. This isn't hard to follow; but the types of policies and premiums are.

The amount of insurance you need is based upon how large of a premium you can afford, as well as the dollar amount your beneficiary will need.

If you're relatively young, have a family, consider how much money it'll take to raise your family in the life-style they're accustomed to, keeping up with inflation and the cost of education. Let's assume you're the head of a household, age 25, with a wife and two children, ages three and five. Until the children reach the age of 18, how much will it cost to pay rent or mortgage payments each year, feed and clothe the children and spouse, provide for other living expenses, and eventually their education?

This should give you a ballpark figure as to the anticipated life insurance you need.

The annual premium can be very large or moderate. This will depend upon the type of policy. There are basically three types: term, straight life and endowment.

First, an examination of term insurance. It has no cash value; you can't borrow against it; it's not an investment, it's strictly insurance. You insure yourself for a term in life. Five years is the average. Some policies are for one year. Most policies permit you to renew it each year until you reach a certain age, about 65. By then you really don't care, because your children are grown and gone.

This type of insurance costs less than the rest. At a younger age you pay less each year, and as you grow older the premium increases. The purpose of this type of insurance is to have sufficient coverage in the event death occurs and you can't provide for your family. You *can* decrease the amount of benefit as the need for coverage decreases.

PRODUCTS AND SERVICES

So-called Straight or Whole Life Insurance is the most costly of all. Most salespersons recommend this insurance. They make more money, and you pay a higher premium.

Besides providing life insurance, this type of policy is also a savings plan but not a very good one. Your policy will have a "cash value." The more years you pay premiums into the plan, the larger your cash value is. You can *cash out* anytime.

Unlike term insurance, the Whole Life premium is fixed, and will never change due to your age. If you choose to cash out (after many years) you probably will get back all you paid in plus about a 5% return. Not really a wise investment, at least in today's money market, but you're covered for the face amount of the policy if you die. You can also borrow, at a low interest rate, on the equity of your policy. There's also tax shelters for persons in high interest brackets.

Endowment insurance has very high premiums, and for your beneficiary to collect, in most cases, you have to die within twenty years after taking out the policy. This type plan is also a forced savings, but at a rate far below today's investment market.

If your goal is to save money, use a bank, savings and loan institution, or other investments; life insurance "investments" only make the insurance company rich, not you. If you want to provide your loved ones with insurance, term life costs less.

The Certainty of Life

Someone close to you has died. Your grief and depression are overwhelming. The task of preparing the funeral arrangements is an added burden upon your shoulders during this time. But you search the Yellow Pages and decide upon a funeral home in your community. At the funeral home you're guided to view the caskets. Your guide leads you first to the most gross caskets available, caskets with red metal flake and gaudy designs. As you move further down the line, you see caskets made with the finest pine, lined with top quality silk, and all of the trimmings.

Your departed loved one deserves the finest burial. And without realizing it, you buy the "top of the line."

In a storm of emotion and sentimentality, you even ponder the idea of burying your loved one with his or her diamond rings and other fine jewelry. (Why bury thousands of dollars of jewelry? The odds are better than even that the jewelry will never get as far as the grave.)

Then you have to choose the burial place. First you're shown the most expensive, a mausolem. The next best is a "view plot" near the top of the cemetery grounds, or near a cross or other worship monument. For the dear departed, the location has absolutely no meaning, but for you it becomes obsessive.

The arrangements are finally made, you attend the funeral, and years later you look back and wonder about your actions during that most difficult time.

* * *

If death occurs, you may be entitled to some benefits. If the deceased was covered by social security benefits, $225 is set aside for funeral expenses. And survivors may also receive one hundred to several hundred dollars a month. Medicare benefits may provide payment of final medical bills.

PRODUCTS AND SERVICES

If the deceased was a veteran, up to $400 can be paid by the United States for a funeral, or internment without cost may be available at a United States National Cemetery.

Some union or employer pension funds may also help in the funeral costs, and also consider fraternal orders. If death was caused by the decedant's employment, Worker's Compensation benefits may also be given.

Nearly five hundred people die everyday in the State of California alone. There is a certainty to life, and that is someday we'll all die.

When a loved one has departed, you must arrange a funeral. It's a sad, necessary task. Don't allow extreme pressure and grief to cause you to make unnecessary, expensive decisions.

BOOK III
HOW TO PROTECT YOUR PROPERTY

I
BURGLARY
Introduction To Burglary

According to the last National Crime Clock, a burglary occurred every thirty-three seconds of every day. Almost two a minute, nearly three thousand a day.

FLASH! As of this very moment, the statistics have drastically changed. A burglary is now committed every ten seconds of every hour of every day. A 300+% increase. There are 3,153,600 burglaries per year.

The annual loss to victims of a burglary nationally is $400,000,000. Four out of every five victims *never* recover their stolen property and aren't insured.

The trend today is that many women work, and they're away from home during the daytime. Thus, more break-ins now occur during daylight hours when no one is at home.

Today, there is a new crop of burglars, the neighborhood teenager. They're amateurs who often commit the crimes seeking thrills. Others burglarize to support drug habits. In San Francisco, the police maintain a controversial squad that arrests drug addicts. They're reported goal is lowering the high burglary rate in that city.

Only 20% of the juveniles are ever jailed when caught. Most often they're put on probation and are back on the streets burglarizing. The Justice Department, as well as the Stanford University studies recently conducted, disclosed that more than half of the burglars entered the victims' homes without using any force. Two out of every three burglars easily entered the victims' homes by opening unlocked doors and windows, or found keys "hidden" in the usual places.

Some insurance companies won't pay a claim unless forcible entry was made: a broken door, a broken window, et cetera.

Since, according to studies, about half of all actual burglaries are not reported to the police, mainly because no insurance is involved, the more than three million reported burglaries each year may actually number 6,000,000.

Burglary is a safe, easy, and profitable crime to commit. You can prevent this from happening to you by following the advice given throughout this chapter. The burglar is your enemy; make his job tough. Don't waste time in helping to catch him ....

Every Second Counts

Every second the criminal has to escape, the chances of apprehending him or recovering your property diminishes.

In reality, the chances of apprehension or recovery are slim.

Even so, when you become the victim of a crime just committed or just discovered, immediately call the police.

Calling the police *immediately* means calling before you call friends or relatives (believe it or not, this is a frequent mistake for some people in panic). Check to see what is missing after you call the police. And your insurance company can wait now, too. Call the police immediately.

The police department tells us that "Every Second Counts," but there is such a thing as priorities and availablity.

Let's take a typical scene: it's dark out; you hear what you think is a prowler outside attempting to enter your home. You have the police emergency number by your phone; you've done everything you know to do to prepare for this situation.

But after dialling the police emergency number, you can expect one of the following responses:
1. A police officer will answer your call promptly.
2. The phone will ring and ring and ring.
3. The line will be busy, or
4. You will receive this recording:
   "This is the police department emergency
   line—all our phones are busy."

Any moment, that prowler will enter your home—to rob you, rape you, or worse, kill you. Even if you're lucky enough to get through to the police department, could they arrive in time to protect you?

The police tell us, "Please be patient."

With all due respect to the police department and other law enforcement bureaus—city, state, and federal budgets produce police departments that are extremely undermanned; they can't respond to every crime immediately. Sometimes it can be hours before a police officer will arrive.

Response time to answer a cry of help is based primarily upon two things: you guessed it—Priority and Availability.

A high-ranking police officer recently told us about a prestigious state official who tried desperately to reach the police emergency board, but out of desperation telephoned our police officer friend at home, and advised him that a prowler was attempting to break in his house.

The police officer called a private line into the police board and requested a patrol car to be dispatched. He was told, "We have 16 emergency calls waiting for an 'officer available' and three divisions in the area have dispatched officers to assist us."

This is so often the case. Calls are so backlogged, many times when the police arrive on the scene, not only is the criminal gone, but so is the victim.

Availablity. Priority.

If you have ever the occasion to monitor police emergency calls on the radio, you'll notice that the familiar loud party, or person blocking a driveway are about the bottom line of the priority calls.

The family or business dispute calls, which often result in physical situations, are also very low on the priority scale.

"Attention, all units: officer needs help ... code three ...."

This type of call is among the most important, and it's not unusual to find ten to thirty police officers on the scene within moments of the call.

An attempted rape or rape in progress, or a shooting in progress, and other such real emergency calls also receive code three responses.

So what is code three? It means to get to the scene as fast as possible, red lights and siren *on*. This response is dangerous—both to the police officers and others in the street, and, unlike on T.V., this is reserved primarily for life and death situations.

Subsequent to code three is code two, and it means to proceed with red light on, *no siren*, and not to run traffic signals, unless it's safe to do so. This response is for important but less serious crimes.

Any other police dispatch is routine, meaning arrive as soon as possible without violating any traffic regulations. Often the call comes out: "Continue to patrol and proceed to..." which means, if they observe another crime in progress or a problem, they're to take care of the observable crime first, then respond to their call.

Frequently, the dispatcher reads three calls at once, so that by the time they have reached the third call, it has been an hour or more since it was dispatched.

Response time to traffic accidents can be very slow. In the City of Los Angeles, traffic units are responsible for handling the investigation of major street accidents. When all units are busy, a patrol car is sent. Unless the call is "Ambulance-Traffic," which means a serious injury is involved, the wait for police can be ten to forty minutes—or more.

"Officer Involved" or "CPD" (City Property Damage) type calls receive a quicker response for police after a traffic accident.

An armed robbery call receives priority, if reported by a person, rather than a silent alarm. An armed robbery in progress is often a code two or code three. If it comes in as a silent alarm, the call is a semi-priority—mainly because so many silent alarms are false alarms.

There *is* an exception. A bank robbery call, regardless if made by a person or an alarm, in most cases, will quickly get a response by a code three with an FBI follow-up.

What about burglar alarms? The old fashion ringers, like those on a business establishment, can ring and ring and will go almost completely ignored. Patrol cars often pass by them unchecked; if it's reported to the police, the response is not fast, and again, this type of alarm is many times a false alarm.

If you discovered your home or business had been burglarized, and the criminal has long been gone, how long will it take the police to arrive? Usually, the fastest response is 20 minutes. The longest? It can be hours, depending upon where you live.

In most cases, no fingerprints are lifted, so don't get upset when you don't see the officer take out his little black kit and begin dusting your windowsills.

What are your chances of recovery after being burglarized? Little or none. If your property cannot be identified, in most cases, forget about it.

Record Your Valuable Property

Someday you may be a victim of a burglary, a robbery, a fire, or other type of loss of some valuable possessions.

If you have the proper insurance you may recover part, if not all, of the monetary value of your loss. If your loss is over $100 or more, you can declare a Casualty Loss on your Federal Income Tax, but only for any amounts not reimbursed by insurance. For businesses: deduct all the loss; for individuals: your deduction is all after the first $100.

Now remember that both your insurance company and the Internal Revenue Service can demand evidence of your claim. Before the loss, take the time to inventory and log your valuable property. Then, if you have a claim, you'll find the time was well spent.

How do you make your Personal Property Log?

Fully describe the valuable items, listing the *make, description, serial number, year purchased*, and the *estimated value*. This information will establish your claim, and assist the police in recovering your property.

Your list should include:

Jewelry	Coins & Stamps
Televisions	All Kitchen Appliances
Stereo Equipment	Power Tools
Radios	Special Equipment
Video Tape Recorders	Bicycles
	Guns
Typewriters	Sterling Silver
Cameras	Sewing Machine

And any other items of value, whether it's covered by insurance or not. Remember—even if uninsured, your loss can be deducted on your income tax.

Next, record miscellaneous items on another list. They should be items of value. Describe them and estimate the year you purchased them and the retail amount you paid.

"Oh, my God, that'll take me several weekends to do!" you're probably saying. Well, no, it won't take as long as you think; get the whole family in on the project, and the time will zip by.

But if you have a portable tape recorder, it may be easier—and faster—to just go room to room, recording in your own voice, the list of items not identified by a serial number:

"In the den there is an eight-foot modern period leather couch, purchased in 1979 for $750; a hand-carved six-foot glass top coffee table purchased in 1974 for $375; an antique grandfather clock from England, circa 1870, a gift from my parents, and valued at $1,000."

And so on.

Proof of ownership can be developed by taking pictures of your valuables.

An inexpensive way is to use an 8mm movie camera. If it has still-frame capabilities, take one frame on each item and/or room area, and you'll have hundreds of pictures on one roll of film!

For you fortunate readers with home video camera outfits, you can record every object in your home on one cassette of video tape.

Another choice: with a 35mm camera, take slide pictures (they cost less than print film) of your valuables. If you have only an instamatic, take wide shots of your valuables inside your home.

Now it's impossible for anyone to later recall each and every item they own. When you're filming your valuables, with movie, still, or video cameras, open the closets and drawers. There are valuables behind them that can be worth hundreds or even thousands of dollars. Take note of your clothing: "One blue, two-piece wool suit, purchased in 1980 for $215."

Don't forget all those precious tools and things in the tool shed and garage.

Most homeowner's insurance has limits. Limits on the amount the insurance company will repay you for your loss of jewelry, money, bonds, other negotiables, furs, watches, and fine arts. Record all these items. Get appraisals. You can purchase additional "all risk" insurance for an additional premium. Jewelry is the most expensive (about $27 per $1,000 value), and for many people the premium for all risk coverage is prohibitive. So photographs of your jewelry—good, clear close-ups—are extremely important for identification if the police later recover them.

If you have a bank safety deposit box, put your Personal Property Log and film in it, not in the house. The house could burn down.

If you don't have a safety deposit box or a fire resistant safe, give the list and film to a close relative to hold for you in their safety deposit box or safe.

It's also wise to spend the money on an inexpensive metal engraver, a vibrating tool that etches anything you wish to write onto metal. They cost about $10.00 or less in many building supply stores. Engrave your driver's license number on the T.V. and other valuables. Don't use your social security number. The police cannot trace you with that. If you don't have a driver's license, put your name and telephone number, with area code, on the item. Of course this is futile if you move frequently.

Keep a list of your credit cards by *name of card, account number*, and *the name the account is under*. Keep one copy of this at home in case they are lost or stolen. Also keep a copy at home of all your motor vehicles by description—license number and vehicle identification number most importantly—in case the vehicle is stolen.

If your car is new, contact the dealer it was originally purchased from. He can tell you the code number of your car key(s). Keep that number in your purse or wallet. If you lose the keys, any locksmith can replace it by the code number, saving you considerable cost.

A final note on your homeowner's insurance policy:

If your policy covers *actual value at time of loss*, which many insurance policies do cover, and you have a loss, you're in trouble. For example: if two years ago you purchased a home video tape recorder for $900, its *depreciated value (actual value*, in insurance language) would now be about $375. That's all the insurance company would pay you for your loss.

On the other hand, if your policy says *replacement value*, then they would pay you *today's* cost to replace it. Beware, though. Some insurance policies claim to offer replacement value, but there is a loop hole. They limit the replacement value to a maximum of *400% of actual value*, or four times that amount. If your television was found off the T.V. pedestal one evening, a perfectly good set, but five years old, costing $600 back then and $900 today, it would be valued by the insurance company at about $100 *actual value*. Four hundred percent, then, would be $400. Had you a *full replacement value policy*, you would have received $900, today's replacement value, rather than the measley $400. Check your policy.

And if you take the time to document your valuable property, making every effort to protect them, then you deserve to keep them. For years and years and years to come.

Have You Set Out A Welcome Mat For The Burglar?

Today started out as a great family day. You're all together enjoying hamburgers at the fast food place, sitting out on the patio, when a police patrol car parks nearby.

"Code seven," the officer calls in. It's their lunch break.

Unlike most civilians, the police have to be on call at all times, so they keep their radio turned up.

"Unit 10T72, a traffic accident at ...." The monotone

broadcast has an urban entertainment about it, and you listen carefully while stuffing down a palmful of fries. You hear, "Unit 10A35 and any available units—a 459 in progress...."

It's your home address that the urban voice behind police broadcast is reading off; you've watched enough television to know that a *459* is a burglary. Someone's in your house, ripping off your brand new video cassette recorder and your jewelry, and the portable T.V., AM & FM stereo, and maybe even your partially hidden Kruegerands worth thousands of dollars, and probably everything you've ever worked hard for!

Will the police arrive in time? Maybe; maybe not.

* * *

Chances are that you unknowingly, unwillingly *encouraged*, if not downright *invited*, the burglar to your lovely home. It could have been prevented. You could have *discouraged* the burglar and kept your hard earned treasures for years to come.

Most of us, though, are guilty of setting out the welcome mat for burglars. In most parts of America, burglaries occur in the home every ten seconds. Why? Because they're about the easiest and safest of all crimes to commit.

You can actually reduce your chance of being burglarized by following some inexpensive safeguards which should yank that welcome mat right out from under the housebreaker's feet.

We'll start when you first move into that new apartment, house, or condominium, or any kind of residence. The very FIRST THING you want to do is CALL A LOCKSMITH. Have him come out and reset the tumblers in your locks. You don't have to buy an entire lock assembly either; just the tumblers need to be replaced.

Why do this? Well, the former occupant has the same key you have, and perhaps friends and neighbors had keys too—and workmen and employees might have keys to your new "castle." It's safer for you to re-key now than be sorry later. But most people buy fire insurance after they've had a fire and burglary insurance after the burglary; and so on. Don't wait. Prevent costly problems and losses from ever occuring.

Now that you have the locksmith there in your home changing your tumblers, go ahead and have him check the outside doors to SEE IF YOU HAVE PIN-TUMBLER LOCKS. This type offers you the most protection, but, by no means, 100% protection.

If you don't have double cylinder locks, which require keys on the inside of your doors with glass panels, install them immediately. That's assuming, of course, you want to keep intruders out of your home. Some ask: "Why spend the extra money for lock security, when I have homeowner's insurance?" Well, perhaps you do. But have you picked up your policy lately and read it? *Carefully?*

Contractors often take short cuts in home construction to save themselves a few dollars; but their short cuts can produce a tall problem for you, costing you thousands of dollars years after the house was built.

Inspect your exterior doors, which should open outward; they should be constructed with hinges having pins that can't be removed. If they can, the best lock in the world won't keep the burglar out. All he needs to do is take a screwdriver and hammer and knock out the pins, and the door will open.

Slide bolts and chains on the inside of your outer doors are very inexpensive and quite effective. But the burglars' delight is the glass sliding doors or windows. They are easily forced open. The spaciousness of open glass makes your home more attractive. And not just to you; your neighborhood burglar finds them very attractive, too.

Two easy security measures can be taken to prevent entry from your glass sliding doors and windows. The most inexpensive way is to place a piece of wood along the bottom track. A neater, but more costly, method is to have a bolt lock installed.

You can discard any sense of safety with a latched window screen. If you want ventilation from open windows, install window locks which limit the opening to spaces small enough so no one can make an entry.

* * *

"This is unit 10A35—code 4. Suspect's G-O-A."

The police arrived at your home. Their code 4 means that no further assistance is needed; in police language G-O-A means that the suspect was "gone upon arrival." So was your T.V. set, your jewelry, your coin collection, and other treasured, personal property.

Will you ever get them back? Most likely, you won't.

Later, the police tell you, "Burglars entered your residense through the louvered windows." They point out, "This type of window is perhaps the easiest of all to pry open. Had you installed safety locks, you'd still have everything." Even your grandmother's sterling silverware set that she'd left you—its sentimental value is priceless—would still be sitting in the beautiful china shelves in your dining room.

The police officer points out, too, that although being the victim of a burglary is like being hit with a bombshell, you can't stay there in the hole with the idea that another bombshell won't hit in the same place twice. You were an easy rip-off, and the burglar won't hesitate to come over sometime for seconds.

Pointing towards the ladder at the side of your house, the police officer tells you the burglar can come back and use it to gain entrance to your property. And if he needs tools, your garage—left open—gives him a variety of YOUR tools to choose from, to aid him in breaking into your house. Sort of like sending bullets to your enemy isn't it?

Today's "free advice" was costly for you, but you listen while the officer gives you one final tip.

"That shrubbery has overgrown. The lawn furniture and that other stuff stacked against the house affords the burglar concealment as he silently breaks into your house." You decide to get out the clippers and find a better place for the lawn furniture, in hopes that you'll not become a victim again.

You have become victim number 916,938—so far this year. If you and all the other victims before you became an organized group—a coalition—in number, you would be overwhelming.

So quit stalling! Be self-protective! Turn that home with the welcome mat into a castle with a drawbridge—and raise it up—for your own protection!

The Burglar's Thoughts

Lenny cruises your neighborhood, canvassing the area for a good score.

And then he spots your home.

"There we go. That's the one."

After parking down the street, he approaches your home and knocks at your door.

"Okay, if someone answers I'll use the old, 'Does Joe Blow live here?'"

He waits. But you don't answer, so he turns the door handle. If it's open he'll enter, and your valuables will exit forever.

"Locked. That's smart, but it ain't gonna help him. Now, let's see. A few well-placed kicks at the door might do the trick, but, shoot, with the size of this door, I don't know. It's pretty dark outside. I could use the jimmy, and spread the jamb and door apart so the lock bolt can be disengaged. Or I can pry the lock cylinder out of the door. Or let me see...." He could even force entry with a cutting tool, which he has brought along, a small hookend tri-bar. It's concealed in his clothes.

"Shall I be a little daring? With a hammer and punching tool, I can definitely break in. No, too much work. No need to work."

Lenny has so many choices, he can't even decide which one to use.

"Back in the car I've got a power jigsaw, with a metal cutting blade. The door is loosely fitted, and it would sure be easy for me to saw off the door bolt. Let's see, uhhh ... the fixture on the light there could be used to power the saw."

One way or another, if Lenny wants in, he'll get in.

* * *

Most burglars are not professionals. However, they earn their "degree" with experience. In areas like Los Angeles County, a vast number of residential and commercial burglaries are attributed to illegal aliens. Youths are considered among the highest number of burglars, and persons addicted to drugs often burglarize to further support their habit.

As long as you're willing to make their entry easy, they'll be glad to oblige you. Most burglars won't spend a lot of time trying to break into your home or business, unless you have something special they want. Burglary is too easy. If you make it difficult for them, they'll find someone else, and leave you alone. But you're not alone, remember—every 10 seconds, of every day, someone in America is being burglarized.

Keys Are Money

This is a true confession of a different kind of burglar. We'll call him Willie. He pays income taxes on the salary he earns as a part-time parking lot attendant for a well-known restaurant. Willie also moonlights, but doesn't declare his income.

In his own words, Willie tells us:

"Just last night I made a big score. It was slow, and I was kicking back doin' a number, listenin' to the radio, and this guy drives in in this shiny black Mercedes. The guy was about 40 or so, and his old lady was a fox—but she wore a mink.

"He had his mind on one thing: the fox. I knew this would be a long supper. Naturally the dude left his keys in the ignition. I gave the high sign to my partner. Here I was sitting with his whole key ring. Every damn key you can imagine is on the ring: house, glove compartment, trunk, safety deposit box, office, everything.

"We opened the glove compartment. His car registration was in the box, home address and all. It's a good thing the fox wore her mink, because everything else was ours."

* * *

This isn't the only method Willie, or the thousands of others like him, use. They can make a wax image of your keys within minutes, and later duplicate the key. Some thieves even keep a portable key duplicator in the trunk of their car.

If you don't keep your registration in your car, they have two other ways to obtain your address. The simplest is to get it from the Department of Motor Vehicles: your name and address are listed by your vehicle license number, and the normal cost is around $1.00. Another method is for an accomplice to follow you home.

With a copy of your keys, you'll be ripped off whenever the burglar wants to do his job. Besides those valuables in your home and office, whatever you have locked up in your vehicle can easily disappear.

Willie continues, "I go through the glove box, the trunk, and I find all kinds of expensive shit. Money, credit cards, tools, you name it, and I take it. The least you'll find is a new spare tire—that's $20 or more anywhere."

So when was the last time *you* looked into your glove compartment or trunk after parking your car?

"So who's to say the shit was there when he came into the lot, right?"

* * *

Never in your life would you hand over thousands and thousands of your dollars to a stranger. To give a stranger thousands of dollars to hold for even five minutes, until you return, would be unheard of.

But you've done it, time and again, without noticing.

BURGLARY

Remember the last time you parked your car in a public parking lot? You left your car and your keys. How many keys did you leave? Of course you left the ignition key—you have to. But did you also leave your office key? Your safe deposit key? More importantly, did you leave your key to your residence? And what about the key to your glove compartment and the key to your trunk? Did he get those, too? If the answers are "yes," you aren't alone. Most people do the same thing day after day. Or until they get burglarized.

Most parking attendants are honest people. But some are thieves. Others work in conspiracy with criminals. So how are you to know the good guy from the bad guy? No, not the color of his hat—you simply don't take the risk.

A burglar, like Willie, can enter your glove compartment and trunk and take all your valuables. And how many times have you checked them BEFORE you left the parking lot to make sure they were still there? And even if you were to discover the loss in the lot, you haven't a chance in the world of recovering your loss from the parking lot company. Check the ticket you got from the attendant. *We are not responsible for lost articles.* And in court, there's no possible way you can establish that the valuables were actually in your glove compartment or trunk when you gave the attendant your key.

Not long ago, we were instructing our citizen's tear gas class, and had about 500 tear gas weapons in the trunk of the car. We had to make a quick stop in a public parking lot before going to a group training session. Even though we didn't leave the trunk key with the attendant, we told him, "Be careful when you park the car. We have some nitroglycerin in the trunk, so you don't want to accidentally move it."

We never in our lives witnessed a parking attendant handle the car with such ease, and so carefully. Furthermore, we knew the trunk was as safe as Fort Knox, so we felt assured in leaving the car in the lot loaded with valuable, "dangerous" tear gas weapons.

The bottom line: never, ever give the parking attendant anything more than the key to the ignition. Purchase the type of key case that permits you to remove one if you choose to.

The key to your personal security is your *key*. "Hiding" a spare house key under the welcome mat, or other common places, is obvious to thieves. The burglar, robber, or rapist knows where to look, too, and the statistics shows us that they *will* look. Hiding the key is not recommended, for when a criminal enters your house with a key, the severity of the crime is diminished, because he didn't *break* in and enter—he *unlocked* and entered.

Never leave your key in the ignition of your car. Not only are you asking for trouble, but guess who may not pay your claim for your stolen car when they establish that the key was left in the ignition? That's right, your friendly insurance company.

If you keep your car in the garage, great—but keep the door closed and locked, or it's only great for the burglar.

When your local newspaper carries a social item like, "Mr. and Mrs. John Adams of Encino are going on a three week cruise to Mexico," you can bet when they get back they'll find they've been conveniently ripped off. This is called advertising your absence.

Another advertisement to the thief: when you leave packages and luggage and other valuable items openly exposed inside your car, you're promoting theft by quick parking lot criminals who can enter your car with an ice pick, start it, and leave, or open the glove compartment or the trunk within minutes without keys.

Some criminals have access to master keys which open any car; others use burglar's tools, and they're masters at their trade. No car or other lock is totally safe from an expert determined to enter. But if you give them access to your keys, the job is fast and easy.

A *warning to physicians*: those medical emblems on your

car, or that personalized license plate, DOCTOR 1, can cause problems. Drug addicts, as you may know, will pry open your car trunk in search of your medical bag. If you advertise your profession, your trunk can be entered. Or worse yet, the addict may wait by your vehicle and attack you as you enter, taking your medical bag *and* your money.

One last thing: don't transfer valuables into the trunk where anyone can see you doing it. Not long ago, one of our clients put the night's cash receipts—several thousand dollars—into his car. Our investigation later determined that an employee had seen him make the transfer from business to car. The employee stole the car, took the cash, and to add a dash of maliciousness to the whole thing, burned the car to the pavement.

Treat yourself to more personal security. Do it by treating your keys like money—because keys *are* money.

Routines and Answering Machines

If you maintain a routine pattern, when you *are* away from home, the burglar will know when to rip you off.

Many of us own answering machines; our voice announces to the caller that we are away from the house, and when we're expected back:

"Sorry you missed us, we're away for the day, leave your message when you hear the beep tone, and we'll get back to you tomorrow."

This is an open invitation to the burglar to empty your house. Your recorded message should have said:

"We cannot answer the telephone now; soon we will get back to you; when you hear the beep tone, leave your message, and we'll return the call first chance we have."

Burglars often stake out their potential victim's house. They note your standard activities. Each Tuesday, at 11:00 A.M. you go to the beauty parlor and return home at 1:10 P.M. The best time to get you is between these hours. You come home with beautifully coiffured hair—to an empty house.

On school days, you take the children to school, leave the house between 8:10 and 8:20 A.M. each morning. Return home at 8:50 to 8:55 A.M. Pick them up around 3:00 P.M., and return back about a half hour later.

How do you fool the burglar? You don't keep a routine. Regardless of what regular appointments you keep, vary the days and times—don't keep a regular pattern.

School days and hours can't be changed for your security. Car pool with other parents. Don't keep the same schedule. Monday, it's your turn. Tuesday, the others' turn, but continue to change the pattern.

Professional burglars learn your habits. If you're a religious person, they know when you'll be at church or the synagogue. If no member of the family can be home, alert your neighbors to pay particular attention to your residence during this period of time.

If a moving van is moving your personal property, your alert neighbor should immediately call the police. If anyone not authorized or known to your neighbor is on your property while you're away, have them call the police. Watch your neighbors' property while they're away, and have them do the same for you.

You'll also find by changing your routine life is more interesting, too.

II
PROTECTION ON VACATION

Vacation Security Preparation

It's that time again. You're about to take a well-deserved vacation. Perhaps you'll enjoy the tropical weather of Hawaii, or ski the slopes of Taos, New Mexico. If you don't make preparations to protect your home, though, before you leave, you may return to an empty house.

Most burglars prefer to enter homes when it's known that the occupants are out of town. There are many security tips given throughout this book. What follows are other measures to take when you plan to leave on vacation.

Most important of all, leave your residence with the appearance that it's being occupied. Don't completely close all the drapes and shades. This could be a sign that no one is home.

Timing devices cost only a few dollars. For most average residences, about three timers are adequate. The timer is plugged into an electric wall socket, and the appliance or lamp is plugged into the timer. You can set the timer to go off at any hour around the clock.

Plug any two lamps or light fixtures into the timer. Set one to go on while the other is off. For instance, you may want one light to stay on in the kitchen area from 6:00 A.M.

to 1:00 P.M., and the light in the den from 1:00 P.M. to 2:00 A.M. This gives the appearance of lights going on and off at different times. The third timing device should be connected to a 24 hour radio station. Keep it on from 6:00 A.M. to 2:00 A.M. Both the lights and radio must be seen or heard from the area of the front door or windows to be effective. You want to deter the burglar from ever attempting to enter.

If you plan to be away for five or more days, ask the post office to hold your mail. Your newspaper, milk or other deliveries must be cut off during your absence. If you plan to be away for a month or more, make some arrangements to have the lawn cut. An overgrown lawn will alert a would-be burglar that no one is occupying the premises.

If you reside in a small community, it's advised to let the police know that you're out of town. And you might want to even tell them where you can be reached. In larger cities, the police may be too busy to patrol your residence, so if you have a trusted neighbor, leave a set of keys in the event of a fire or other emergency. Let your neighbors know you're out of town and how to contact you. If they see someone on your property, they can call the police.

We don't recommend putting your telephone on "vacation." Most phone companies will temporarily disconnect your phone while you're on vacation. You may save paying a few dollars to Ma Bell, but you're also verifying your would-be burglar's suspicions that you're out of town.

What about a big sounding watchdog, that you don't have to feed, or clean up after, but has a bark that'd scare off a burglar? Well, if you own a cassette tape recorder, a 24 hour cannine can "guard" your residence. Purchase a continuous tape cassette, the kind used for answering machines. They run twenty seconds and forty-five seconds and continue to repeat whatever is on the tape.

Record the barking and growling of any dog, or tape it off a sound effect record. The noise will give the Bad Guy second thoughts about entering your house. This "trick" can be used anytime you're away from home. Just make sure that the barking can be heard from outside the house.

If your garage door is unlocked, or can easily be entered, and the burglar knows that you're out of town, he may drive his own car into your garage, close the door, and make his entry through the garage entrance door. He can easily enter the premises, load his car or van with your belongings, and drive away. Even if your garage door is unlocked, if you own two vehicles, or keep one at home, park it (and lock it) in the driveway, to block the garage door.

Enjoy your vacation.

Your Airline Reservation Doesn't Guarantee You A Seat

Nearly 200,000 persons last year had "guaranteed plane reservations" and were "bumped from the flight." This is the most frequent complaint against airlines: overbooking.

When you're "bumped" from the flight, don't take the meaning literally. You won't get pushed off the plane. You just won't end up on the flight you thought you would.

The Civil Aeronautics Board is a federal agency that regulates the airlines. They also permit the carrier to overbook. It's economical.

Can this affect you? Perhaps a relative passed away. You booked a flight to the East coast. The funeral was scheduled at 10:00 A.M. tomorrow and your flight would arrive at 6:00 A.M.

You didn't sleep that night, packed your things, and drove to the airport. You had "confirmed" reservations for the next departure, but when you checked in, you're told "sorry, we have no space available." And there's no other flights that would get you there in time for the funeral.

There's a chance that you could get on board if one of the

other passengers would volunteer to give up their seat. The airline will ask, but odds are against you. The airline will offer some compensation for the inconvenience of giving up the passenger seat. Few people, however, will volunteer.

You have now been denied boarding, and are entitled to on-the-spot payment for compensation. The amount: no less than $37.50 and no more than $200.

If the airline can't board you on any other flight, theirs or a competitor's, where you can reach your destination within two hours of your initially scheduled arrival (four hours for international flights), then the compensation doubles, not less than $75 or more than $400. The amount paid depends upon the cost of your ticket. The "compensation" may not be equal to the damages caused you by the airline. If this is the case, don't cash the check given you; you may wish to file a lawsuit. Consult an attorney for legal advice.

Denied boarding compensation doesn't apply if you didn't meet the airline's deadline for purchasing your ticket, or you didn't check in at least ten minutes (or more) before the plane's scheduled departure time. It doesn't apply if the airlines substitute a smaller plane than what you're scheduled to take due to mechanical reasons, or if you were bumped to replace an official traveling on emergency government business. This rarely occurs, however.

Even though they aren't required to do so, many airlines will pay, in addition to compensation, for your long distance telephone calls and meals, if the delay is four hours or more.

Besides being "bumped" from an airline, the second most common complaint is lost baggage....

PROTECTION ON VACATION

Missing Luggage Syndrome

Most major airlines lose over one million dollars a year in lost luggage. Many frequent travelers experience this inconvenience. Arriving at the airport just minutes prior to the flight, you risk having your luggage placed on a plane other than the one you're boarding. Arriving early does not guarantee the safe and prompt delivery, either, but you have a better chance.

Under a new law, all air carriers require that you put your name and address on the outside of your luggage. They also provide tags. To save time, tag your luggage before you arrive at the airport. Carefully pack your luggage. Most locks can easily be picked, but locked luggage provides better security. Many suitcases look alike. To easily spot yours, put an unusual decal or marking on both sides. Fragile items are not insured for damage by the airlines. All vital and important belongings, medicine, irreplaceable items, jewelry, money, negotiable papers, art or other such items should be in your carry-on luggage. Under CAB regulations, airlines are not responsible for such valuables.

The standard liability per passenger is $750. This is based upon the depreciated value. A $300 suit purchased a year ago has a depreciated value of about $100. The airline, not you, determines its depreciated value. You won't get the cost to replace the lost item. Read the notice on your ticket.

If you're carrying more than $750 (depreciated value), you can increase the airline's liability by paying "excess valuation" when you check in. This "insurance" costs 10¢ per $100.

If your luggage is damaged in flight, from the time you give it to the airlines until you take it from the baggage room, most airlines will usually pay to repair or replace it.

When your luggage is lost, report it at once. If you took sporting equipment such as skis, the airlines will usually pay you the cost to rent equipment. If your clothing was lost, the airlines will pay, in most cases, 50% of the actual cost to buy new clothes, or even 100% of the cost, if the luggage is later found and you turn in your old clothes.

You have 45 days to file a signed lost report. But you must first make some report as soon as the luggage is missing. When found, airlines will deliver your luggage to you, but often you have to insist that they do so.

Direct any airline complaints to: Civil Aeronautics Board. The local address is listed in your telephone directory.

Travelers' Intruder Alarms

When you're out of town and check into a hotel or motel, do you feel secure?

It's time to retire for the night. You put out the Do Not Disturb sign, bolt the lock, and shut off the light. How safe will your valuables be? How safe will *you* be?

First of all, you aren't the first person to occupy this room. It's rented to hundreds of people each year. The lock is not changed every time a new guest checks in either. Many professional burglars rent rooms solely for the purpose of duplicating the key for future entry. Who else has a key to your room? The management, room clerk, bell-hops, maintenance people, room maids, and so on.

Lesson #1: if you're carrying valuables, check them into the hotel safe deposit box or safe. This will protect your valuables. But what about you?

Most seasonal travelers, for safety's sake, carry traveler's checks. Today, some of the more security-conscious people are packing in their luggage a palm-size intruder alarm. Even though the alarm is portable, it can also be used on a permanent basis for apartments or homes. It's powered by a nine-volt battery, and can last an estimated

500 hours. The alarm is activated as soon as the doorknob is touched. We know of three manufacturers of these alarms, which sell for around $25-$30.

1. The *Startler K6700* (Regalware Incorporated, Kewaskum, Wisconsin), is sensitive enough to detect a gloved hand on the doorknob.

2. *Entry/Alert EA1* (Entry/Alert Company, Carlsbad, California).

3. The *Door Sentry DRSDC* (P. R. Mallor and Company, Indianapolis); alarms respond only to a bare-handed contact.

The *Entry/Alarm* will give off a loud alarm lasting 45 seconds or more after the doorknob is touched. The other two cut off the alarm once the hand is removed from the doorknob.

What happens when a drunk accidentally touches your doorknob in the middle of the morning? Well, you'll be awakened. But prepared. In most situations, it's best to have a false alarm than no alarm at all.

Crime Against Tourists: Vacationing in Paradise

Over three million persons visit "paradise island" each year. This is where heaven and earth are one, where breathtaking scenery and exquisite beaches overwhelm the "mainlander." You're on vacation and enjoying the tropical lifestyle of Hawaii.

Some call these islands a "mythical heaven"; perhaps this is why the popular television program, *Fantasy Island*, is filmed here. In real-life, everyone can play out their fantasies in this paradise. But for many, their holiday will be marred. They become victims of burglary, robbery, assault, rape, and sometimes murder. The information

you'll be reading can be applied to almost any city or state in America *you* may visit.

Co-author, Milo Speriglio, went to the islands to investigate. His mission was known only to the Governor of Hawaii and other state and local officials.

* * *

I stare at the darkened stage in silence, a *mai tai*, a popular Polynesian drink, in hand, spotlights on a native islander attired in a stunning flowered shirt, white pants and shoes. Our singing host has one thing in common with all of us: we're all wearing the badge of the islands, a *lei*. Kiwini greets us with a big *Aloha*, and in the tradition of the island, we reply, *Aloha*. He walks slowly to our table, microphone in hand, and asks, "Where is home?"

We answer proudly, "California."

He moves from table to table asking, "Where is home?" and the audience chants back, "New York ... Japan ... Dayton, Ohio ... Switzerland ... Los Angeles ... Boston ... Texas ... Canada ... Corry, Pennsylvania ... Tennessee ... Mexico City ...." The Lu'au, the Hawaiian feast, begins.

People come from all over the world to visit Hawaii. The Hawaiian Visitor's Bureau reports 3,036,033 tourists came in 1980. Less than the prior year, the down trend continued in 1981. Tourism became a serious enterprise for the islands shortly after statehood, and developed into a leading industry in recent years. The national press notices of crimes committed on tourists is one reason for the decrease in visitors. The higher cost of airfare is another.

My wife and I arrived in Hawaii on *Poaha, Iulai, Ekahi Enco* (Thursday, July 16th); Kauai, the "Garden Isle," was to be my headquarters. I stayed at the Kauai Surf Hotel, in Kauai, on the oldest and forth largest island. Where the movie *South Pacific* was filmed.

While undercover, I learned a little of the islands' language. It made paradise more realistic. *Aloha* is the first word you learn; it has many meanings: hello, welcome, love, affection, and good-bye. If you become a victim of a holdup, though, don't expect the criminal to say *Aloha;* more likely he'll say, "good-bye sucker." You'll never find a warning in the travel brochure that you may become a victim of a crime on "paradise island." Once you arrive you may be warned, but like most tourists, you probably won't pay much attention.

If you smoke cigarettes, you're warned by their advertisements, and on the product. "WARNING: The Surgeon General Has Determined That Cigarette Smoking Is Dangerous To Your Health." All are warned, but many still smoke! When we arrived in Hawaii, I observed several bumper stickers on rent-a-cars, which read, *"The Driver Has Been Warned Not To Leave Valuables In The Car."* What did I see inside? Cameras, purses, clothing, even a key in the ignition. Soon, these valuables will become the property of the burglar.

Warnings serve a purpose, unless they're ignored. Inside our hotel, we noticed an almost poster size *warning* on the door, and on the sliding glass door to the patio there was a large silouette of Sherlock Holmes, and behind it, in large print, "ELEMENTARY... there are some people even in Hawaii who simply can't resist temptation ...." They concisely listed safeguards of security protection, which are covered more extensively in this chapter. The hotel also pointed out that Hawaiian law limits their liability.

Deputy Chief Ray Duvachlle was in charge of the Kauai Police Department during this interview.

"Should tourists be concerned about crime on the island?" I asked.

"You have to be alert no matter where you are. And Hawaii is no exception." He said crime is not as bad as it was, and most crimes are thefts from vehicles. He admitted, "About one-third of our reported rapes, a tourist is the victim."

Crime against tourists was a problem throughout the

state. Our local radio station reported crimes committed everywhere—even in our hotel, suspects were arrested for assault, robberies and muggings.

Roy K. Hiram is the Chief of Police of Kauai, and the author of the following article which appears weekly in the *Kauai Beach Press,* a division of the *Hawaii Press Newspapers.* Bill Raul, the publisher, has granted us permission to reprint it:

Dear Visitors: The men and women of the Kauai Police Department would like to take this means of extending our concern for the need for each of you to be aware of the need in properly securing your possessions to insure a well-earned, safe and memorable vacation free from the problems of crime. To insure this, we would highly recommend that you abide by the listed suggestions below. These suggestions do not guarantee that you will not become a victim of a crime, but it will reduce the possibilities.

HOTEL: Store valuables within the safe deposit or security boxes that are regularly available. Upon leaving the hotel room, either for a tour or to sunbathe/swim, properly secure all exterior doors and windows to the room, and leave room keys at the front desk.

SIGHTSEEING TRIPS: Don't take unnecessary valuables along with you. Bring only the equipment and items necessary to make your tour pleasant. If renting a vehicle, don't leave valuables unattended within your vehicle. Remove all valuables from your vehicle/bus whenever you leave same, even for a short time. Never leave valuables unattended on the beach or in your vehicle when going out for a swim. Have someone watch or remain with your possessions.

CAMPING: Erect tents only in those illuminated areas ... close to others that may already be there. Don't leave tent or equipment unattended. Always have someone remain at the campsite to provide security.

The Chief also advised to let someone know your destination and date of return when hiking; two Californian's were recently murdered on Kauai while jogging. His final word was: "Remember, should a crime occur, *DO NOT TOUCH, MOVE OR CLEAN* anything until an officer has been called and an investigation has been completed."

The warning was clear, and was signed, "With Sincere Concern." The Oahu edition of *This Week*, a major tourist guide, has a section called "Security Tips for Tourists," and here is one of their recommendations: travel in groups when sightseeing in remote areas of Oahu, particularly the Waianae coast and Waimanalo. They strongly urge that you don't camp in these areas because of the high crime rate. The *Hawaii Tourist News* advised, "Tourists' autos are tempting targets at known scenic attractions, and a locked trunk or glove compartment is no hinderance to an accomplished thief."

Crime against tourists is a major concern to both government and business in Hawaii. The *Waikiki Beach Press,* another tourist newspaper, headlines a warning, "*ALOHA-TAKE CARE.*" They advise, "The combination of sun, surf, scenery and the special people of Hawaii with their aloha spirit, make a stay in the state a memorable one for all who make themselves aware of certain safety precautions." Tourists are warned not to walk darkened streets and alleyways late at night or in the morning hours, and not to travel alone. Other visitor's newspapers, such as the *Guide to Oahu,* the *Waikiki Bulletin* and the *Paradise News* didn't offer their readers any security advice.

Crime in paradise will decrease. The Honolulu City Prosecutor, Charles F. Marsland, Jr., was the first elected Prosecuting attorney in 50 years. He was sworn into duty on January 2, 1981. In the first three months, there was a reduction in crime for the first time in decades.

"The sad fact is," Mr. Marsland stated, "that one has to be cautious in any major city in the U.S. today—including Honolulu."

Mr. Marsland went on to say that tourists are victimized because the "criminals know that tourists—especially visitors from Japan—often carry large sums of cash." The problem was conglomerated by the fact that victimized visitors have been reluctant—or not able—to come back to the islands to testify in court.

In cooperation with the Waikiki Improvement Association and some airlines and hotels, Mr. Marsland has succeeded in making the tourist less "attractive" to the crooks.

In the first six months of 1981, Mr. Marsland's office got 40 victims and witnesses to come back to Hawaii—at no charge to them—to testify at some 21 trials. The result? 19 convictions; a 90% conviction rate in crimes against tourists.

Crime on the islands is not restricted to tourists. I was amused to learn that during my investigation, cattle rustling was wide-spread, particularly on Hilo, the big island. *Crime Stoppers,* an anti-crime organization we'll soon discuss, offered a $1,000 reward for the rustlers.

While operating undercover on the islands, I met Kana (not his real name), a youth specializing in luggage burglary. He referred to himself as a surgeon; his "operating" tool was a pocket knife. "Like magic," he said, "I open the lock on the suitcases, take what I want, and go." Kana preferred entering the tourists' luggage while it was lined up waiting for room delivery in the hotel lobby. Within 30 seconds or less, your "locked" luggage can be entered; the thief walks away with your valuables.

When we arrived at the hotel, I discovered we lost the keys to one of our bags. Using the "trade secrets" I had previously learned, I opened the lock with a pen knife. Time lapse: about four minutes. I was no professional, and as the pros would later tell me, "practice makes perfect."

Billy Ervin knows how wide-spread crime is in Hawaii; he's the Chief of Merit Protective Service of Hawaii. This private firm is retained by many insurance companies. He told me, "... Milo, there has been a lot of crime on the islands, especially on the 'big island.' So many burglaries, robberies, purse-snatches, rapes, something goes down every hour ... even the newspapers report it."

My colleague told me, "Employment (on the islands) is limited. We have the *haves* and the *have nots*. He explained that criminals rip off tourists most often when they display jewelry, cash, cameras and other valuables. Carrying "wealth" only makes you a potential target, I learned. *Operation Sting,* as the HPD (Hawaiian Police Department) called it, was an undercover operation where detectives set up store fronts and bought "stolen" items from fences. The operation was a great success, however, most of the recovered items were never redeemed by the victims.

The islands now have a new type of "sting" which helps take the bite out of the criminal. On Kauai, it's called the *Anti-Crime Secret Witness Program*. Ezra Kanoho started the program in 1979. He's the Director of this association, sponsored by the Chamber of Commerce, and is also the Island Manager of the Hawaiian Telephone Company. This program is similar to the *We Tip* program reported in Book I, *How to Protect Your Life*.

Mr. Kanoho told me, "There is cause for concern (about crime), and we have to do something about it—it effects our economy." Here is how the *Secret Witness Program* works: If you have any information regarding a murder, robbery, rape, manslaughter, aggravated assault, grand theft, auto theft or burglary in Kauai, call 245-8640, and your anonymity will be preserved, or write to *Secret Witness,* P.O. Box 1969, Lihue, Hi. 96766.

You'll be given a reward—by certified check—for up to $1,000 or more, depending upon the nature of the crime. The payment arrangement will be discreet and you'll remain anonymous. One of the most recent publicized crimes on Kauai involved a California attorney and his wife, murdered while jogging. *Secret Witness* offered a $10,000 reward; so far, no takers.

In early 1981, a similar program was adopted throughout the rest of the Hawaiian islands; it's called *Crime Stoppers*.

It's a project of the Chamber of Commerce of Hawaii, the Police Departments and the news media. They, too, offer rewards for up to $1,000 for information leading to the arrest and indictment of any criminal. The telephone numbers are: Hawaii, 955-8300; Hilo, 961-8300; Kona, 323-3300; Maui, 242-6966.

During my investigation on the islands, there was a case of mistaken identity. A tipster identified a 21-year-old Hawaiian who was arrested and charged for attempted rape and burglary. He was released when another man, age 26, was arrested for this same crime. Police later said, "They (suspects) could have been twins." Rarely will a case of this nature happen. These anti-crime programs are an invaluable aid to the justice system.

My next "informant" was on the wrong side of the law, a criminal, like so many others interviewed during my investigation. I'll call him Pila Kamilka (Bill Smith in English). He's a young native, a high school dropout, and has no steady job.

"I don't give a shit about the tourist, man. They show me their gold, and it's mine—I'll take it. If I was on the mainland, I wouldn't be so game to take them, but here we know that they won't come back and testify in court. All they want to do is get their rich asses back home ...."

Who should know more about crime than the Attorney General? In Hawaii he is Pany S. Hong. My first question to him was, "What is your attitude about crimes committed upon tourists?" "I'll be honest about it," he answered. "I haven't given it a thought. I *am* concerned about crime in Hawaii."

I asked, "Mr. Attorney General, are you not the state's attorney in matters of crime?"

Tourism was the state's major source of revenue, and crime was an issue. The Attorney General told me crime against tourists has never crossed his mind. I learned he was a member of the Public Prosecutor's Association, and that he and his office are a key figure in influencing state legislation—civil and *criminal*.

The Governor, according to a top aid, was very concerned about crime. I learned that just two weeks prior to my investigation the Governor of Hawaii had told the Attorney General to help find a solution to crime in Hawaii.

The Attorney General of California is the chief law enforcement officer of the state. Your authors are certified by him to instruct and license persons to carry tear gas weapons for self-defense.

I asked the Attorney General of Hawaii for his opinion of tear gas. "I am not certain," he said. "I did not really follow Mace. All I know is that it is not legal in Honolulu, however, they are trying to make it legal."

That night, like each in the past, I watched the local television news. The subject was tear gas; should it be legal in Honolulu?

My undercover role was exposed. "Pat," I called to my wife, "I'm on T.V." The film clip taken from a prior network television program showed me as a detective, demonstrating tear gas.

The next morning, I interviewed Phil Yamagata, the owner of Yoshimura Gun Store. He's an angry man. "When I was a young person, owning a gun was rare, but today it is necessary for everyone to own a gun."

Phil is a law and order man. He believes that repeaters must be jailed. Even on Hawaii, when crime sprees are reported on the news, he said gun sales go up. He's angered about the new gun law, House Bill 293-Act 239. To buy a gun on Kauai you must now wait no less than ten days and no more than 15 days. You must be fingerprinted and photographed. Unlike other cities, this restriction applies not only to handguns but to rifles and shotguns. He told me he'll now sell tear gas to civilians—something he only sold to the police before. In Kauai, tear gas is legal.

Francis Keala, Chief of the Honolulu Police Department, said, "Hotel crime is a major problem in the City and County of Honolulu. It cannot be eliminated entirely in the foreseeable future, but it can be reduced substantially through a concerted and continuous effort by you and the Police Department."

The Chief warned, "Beware of prostitutes. They've been known to steal from men whom they have enticed." He further advised, "Some prostitutes work in teams on the street to pick pockets; others go into hotels where they gain entry to rooms and rob the tenants."

Is rape a problem in Honolulu?

The Chief said, "Rapes and assaults are problems of increasing concern in Waikiki." He advised using caution in the following areas:

    1) Restrooms open to the public.
    2) Parking lots and garages.
    3) Beaches at night.
    4) Stairwells.
    5) Alleys and isolated walkways.
    6) Public parks at night.
    7) School campuses and playgrounds at night.

His final advise was the same as nearly every law enforcement official interviewed, "Avoid going out alone at night."

* * *

As we've said, the largest complaint received by airlines is lost baggage. When you add the cost of the items you bring when going on vacation, it may well exceed the $750 liability which most airlines have. You can buy extra insurance, but there is a catch. You must declare excessive value before you turn it over to the porter, and an inventory must be made at the ticket counter.

But if your airline didn't lose your valuables, you may lose them in your hotel room. The hotel rooms we stayed at in Hawaii were typical of most good hotels anywhere. The security appeared to be better than in most homes.

The doors were solid; two inches thick. They had an

expensive deadbolt lock. By pressing a switch on the lock no one could enter—even with a key. There were chain locks. The only disadvantage of the switch and chain locks was that someone must remain inside to lock them.

The key that fits your hotel room has been used by hundreds of other guests, and they could have duplicated it. Nearly every hotel has a policy that the housekeeper must keep the door open when cleaning the room. After finishing breakfast on the first day of my trip, I returned to the room, and to my amazement, I discovered how easy it was to burglarize any room in the hotel.

"Aloha," I said to the housekeeper, whom I never had seen before.

"Aloha, sir," she replied.

I picked up my tape recorder and camera, and walked out of the room. Anyone could have done the same.

Hotels should require the maid to see the room key before permitting anyone to enter the open door. It's impossible not to leave some valuables, such as clothes, in your room, but put all cash, traveler's checks, credit cards, jewelry, and other small valuables in the hotel safe deposit boxes.

When you visit the island of paradise, and have a complaint or comment, contact the Hawaiian Visitor's Bureau, phone 923-1811, or write HVB, 2270 Kalakaua Avenue, Suite 801, Honolulu. When in Waikiki, there is a visitor's information program—call 836-6413; for fire-ambulance or police, call 922-2323; 24 hour physician, 536-4346; 24 hour dentist, 521-4555.

I plan to return to Hawaii, and I'll be better prepared to make my vacation safe. It's time to return to the mainland; I'm more tanned and much wiser.

* * *

My investigation in paradise is over. We're aboard a 747, flight number 196, to Los Angeles International Airport. Midway through our journey home, I engage in conversation a man in his early 30's, wearing bermuda shorts. The conversation gets around to "What line are you in?" It turns out he's a Special Agent for the F.B.I. Among his many assignments was to protect President Reagan. On the way home we exchanged "war stories," over several Mai Tai's.

The pilot turned on the "fasten seat belt" sign, and Margaret Fahey, the flight director, announced, "We are preparing for landing at LAX."

Touchdown! The big bird lands smoothly. There's applause heard from first class to coach. I noticed everyone—the stewardesses, stewards, and passengers—were doing the same.

Throughout the whole aircraft's cabin was applause and tears. No, we didn't just avoid a collision, or a forced landing. This was a very special moment. Captain Frank Hart just ended his 37 year flying career, and retired on this very special last flight.

Aloha.

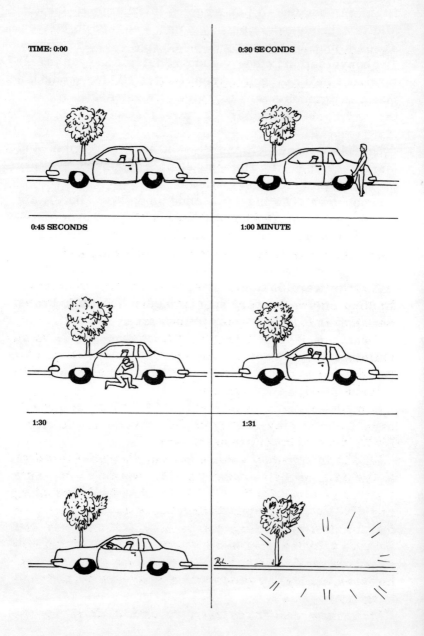

III
AUTO THEFT

Abracadabra: There Go You're Wheels

Auto theft and burglary—like other crimes—is growing. In Los Angeles alone, about 33,000 vehicles will be stolen and another 45,000 burglarized each year.

Every one of us is a target for thieves. If you drive an exotic car or a Volkswagen, someone wants to strip it for its parts, at the least, and, at most, they want to sell it in Tijuana, Mexico as a taxi cab.

Sometimes all they want is *some* of your car. The cassette or eight-track player, or your C.B. unit, or, if you own a BMW, it can be the entire dashboard.

Who buys anti-theft vehicle burglar alarms? According to the dealers, their customers—95% we were told—were victims of a car theft or burglary. For $150, including installation, you can purchase a heavy-duty ignition lock for the steering column and ignition kill-switches that prevent vehicles from being started, or devices that will sound alarms if your ignition key isn't used. These devices, however, are merely minimum security against theft and burglary.

Many new cars have theft devices built in by the

manufacturer. Most common is the steering wheel lock that won't turn without the key.

Alarm systems can be programmed to activate (horn blowing or a ringing alarm) if the door, the hood or trunk is opened without your knowledge. This system usually is accomplished by installing pin switches similar to the type that activate your car's dome light when your door opens.

Some insurance companies, by the way, will offer you a discount if you have some kind of alarm device installed in your car. Regardless of the anti-theft alarm you choose, remember, they have limitations. If a thief wants your car bad enough, he can resort to towing it away. And no matter how much you spend for vehicle security, it won't prevent a theft or rip-off 100% of the time. The best it can offer is possibly deterring the common thief from entry and increasing the chance he'll be caught in the act.

This is a composite of law enforcement officers' personal opinions:

"Alarms are only as good as the person(s) breaking into the vehicle. If the courts handed down stiffer sentences, vehicle crimes would shrink, not grow. But some protection is better than none."

How safe is your vehicle—even with super anti-theft devices? We interviewed a professional car thief, and you'll be surprised by his remarks.

Brad is 21 years old, from a Los Angeles middle-class home, has a job, a high school diploma, and, at this writing, is on trial for three counts of receiving stolen property and two counts of Grand Theft Auto.

How They Get In

Q. How does the professional car thief get inside someone's car?

A. Usually we go in through the windows. About 90% of the time we're in a rush, we cut the rubber out of the back window with a single-edged razor blade. Just pop the window and pull it out. To pick the lock, you go exactly one-half inch above the keyhole with an ice pick and poke it right through the door. Then I hit the lever up, and just bypass the key.

Q. How long does it take to do that?

A. As fast as a key if you know what you're doing.

Q. It's just like a surgeon then?

A. Exactly. And once I'm in, I use a device called a slide hammer which auto body people use to pull out dents.

Q. What do you mean it takes out dents? It's like a suction cup?

A. Sort of. It has an adapter in the end like a drill. It's a long bar with a weight on it and a drill on the other end. I have one at home. Just like the movie with James Caan called *Thief*. The movie shows how to steal *anything*. Especially from safes and houses. All of the same tools apply. He used the slide hammer to break into doors. Like your house door. Instead of putting a key in it, I'd put in a little drill bit, tighten it down and screw it in there just like a drill. Then I just pull it out with one good whack. Then the lock assembly is gone and I can use a screwdriver or anything to turn the latch.

Q. How long does it take you to do that?

A. One whack. Five seconds. I'm "gone in sixty seconds."

Q. Is that the standard amount of time?

A. Well, if it's a Porsche, I have to get it started within 17 seconds or else the alarm goes off. I just pull out the regular key and use a screwdriver. To sell the car all I have to do is buy a new ignition.

Q. How much would a new ignition cost?

A. Porsches are about $110.

Q. What about the Volkswagen? There are a lot of Volkswagens in this country.

A. 1967 and lower will start with anything.

Q. Any key?

A. Any one of these keys would start it. I just put it right in there. There's certain ones you'd have to break, but if I can wiggle my key in there. I can start *any* "V-dub."

Q. Most people don't lock their cars, so all you have to do is just climb in, turn the key, and drive away.

A. Volkswagen should have done something about that. It's just a result of a bad ignition system. The best thing to do to prevent somebody from getting in your car ... well, you can't really prevent anybody from taking the car, but you *can* prevent them from getting away by installing a gas shut-off switch.

Q. What are the various devices to deter someone from getting into the car?

A. From getting in? Motion detector.

Q. Motion detector? Do you know some brand names?

A. *Cliffords.*

Q. How much do they run?

A. A lot of money.

Q. So if you even touch the car, it'll go off?

A. Exactly. They have a little key there, but they still can be beat.

Q. But if somebody leans against the car, it's going to go off?

A. Exactly.

Q. Will it stop once the motion stops?

A. No, it's continuous. People also have silent ones that they carry on them. So if it's touched, the thief doesn't even know it. Those are the tricky ones. If you want to spend a lot of money you can really prevent someone from breaking in. But the average person doesn't know anything about taking a car. I was working the other day and a limousine was stuck on a hill with a dead battery. They asked me if I would give them a jump. I said sure, but it still won't start. So I just started it through the starter. I can start American cars strictly from the starter to the solenoid using just one wire.

Q. Do you get underneath the car?

A. No, you open the hood, find the positive battery cable, put a wire to it, and run it to the starter. Then I touch a

screwdriver to the solenoid and it'll kick over the car.
Q. With a dead battery?
A. With a dead battery! I just bypass the battery. They gave me a $20 tip. It was a pink limousine.
Q. We've talked about how to prevent someone from getting in, but how do you prevent getting it started? What are the various devices available?
A. When you leave your car, take parts off your motor.
Q. Is there something you can do really quick? In a couple of seconds?
A. Take off your distributor cap and put it inside your car trunk. Just unclip it.

How They Make a Stolen Car "Legal"

Q. Let's take the entire process, with a Volkswagen for example.
A. First, I buy a crashed (junked) car. It's legal: pink slip, license plates, everything. Current registration if possible, because I want to take it to the Department of Motor Vehicles to register it. And when I get the new plates, I'm in great shape. So I have my "legal" car. There's a tag underneath the front hood, a tag in the window and one underneath the back seat on the floorboard. First I torch the rear number out. Then take the window tag and hood tag off.
Q. They come right off?
A. Yeah. Drill out the rivets. I put the "legal one" on the street with no markings on it. I take the license plate, the rear numbers, the window number and the body number off the car. I put it back on the street, the police find it, and I watch where it goes. They don't know where it's from because there aren't identifying marks on it. So they just sell it off in an auction. Now, I have all the tags to a legal car, and I have the pink slip. Now I steal myself a Volkswagen, take out its numbers and put yours on it.
Q. When you put the tag under the hood you have to use

certain types of rivets, don't you?
A. The Highway Patrol puts them out.
Q. How do you get them?
A. Well, I just have a few hanging around. Sometimes I steal rivets from Sears.
Q. So, you buy the crashed car, take all of the identification off it, steal another car, and take off all of its identification? Then you throw the identification away on the one you stole, take the good identification off of the crashed car and put the I.D. on the new one? The new car then becomes the crashed one?
A. And it's "legal."
Q. Doesn't the Department of Motor Vehicles know that this one is crashed?
A. No. You don't tell the Department of Motor Vehicles that you crashed a car.
Q. But when you buy a car that's junked, are they allowed to sell it for other than parts?
A. No, you buy it from a private party.
Q. You just buy it from anybody that has a crashed vehicle?
A. Don't go to the junk yard.
Q. It has to be registered still?
A. It's better.
Q. You get rid of the crashed car, and they find this crashed one with nothing on it. You're driving down the street in this "legal" car, registered under your name. There is no way of knowing any differently?
A. I personally have had a car, a Porsche 911, where (the true owner) knew it was his because of the crack on his dashboard. But I had all of the receipts and numbers matching the car, and the person knew it was his, but couldn't do anything about it.
Q. So what do you suggest? If someone ever does see their stolen car on the street, how can they identify it?
A. Initial it.
Q. Where?

A. Anywhere, inside. On the chassis even. Motorcycle gangs' initial everything on their bikes from their pistons to their handlebars. If a biker ever finds anything stolen on someone else's bike, they've had it. You need something identifiable like a special kind of header that you have a receipt for and can identify. A crack in his dashboard doesn't mean anything.

Q. If someone were to take an engraver, go underneath the car and engrave their initials on the chassis, when it's stolen later, all you'd have to do is go under the chassis to prove it was yours?

A. It's proof without a reasonable doubt. It's hard to be caught for things the way the law is this day and age.

Q. How did you get caught?

A. I bought a stolen car to begin with. Then I stole another one, and I re-stole the stolen car.

Q. How many cars did you sell before you got caught?

A. I personally sold fifteen. I made about $7,000 a piece for them. But then we steal them back from you after we sell it to you. Then, nobody has any identification on it.

Q. You steal them back?

A. I just steal them back for parts.

Q. So you make the money selling the car *and* on the parts. How soon after they buy it do you steal it back?

A. Usually a couple of days. The Volkswagen I got caught for had Nevada plates. The people were visiting so we figured they'd be no problem at all. The Department of Motor Vehicles won't even know about it because it was from a different state. The crashed car that I bought was already stolen. I bought it for $100 through the paper. When I did the change and registered it, it came back three months later stolen. The girl told who she bought the good one from, which led to me. I didn't think anything would happen.

When, Where and the Prevention

Q. What is your preference? Day or night?

A. Nightime. It doesn't really matter though. In the daytime I have gone into a Porsche-Audi dealership and walked out with a new Porsche. Just walked in, jumped in it, and drove out the driveway. A brand new car with keys in it. It was in for service. I just walk into the service department with a blue shirt on and don't even get questioned.

Q. What areas where cars are parked would you least like to get involved with?

A. Underground parking lots. Dangerous because the slide hammer makes a bit of noise. That's because of the weight.

Q. Tell the readers what happened to you the time you got into the Porsche in the apartment underground parking.

A. Well, there's a very advanced alarm where the gas shuts off. I had the car started, got in it, drove it up the driveway, and the thing ran out of gas. I looked underneath the dashboard for the little switch to turn it back on. Usually they're underneath, close to the driver. Smart people put it in their trunk. It's better to stake out a car. Watch the people for a day. Run the license to find out whose it is. So anyway, I was driving up the hill when it ran out of gas. So we looked underneath the dashboard to find where the gas shutoff switch would be. We couldn't find it, and all of a sudden the two doors locked and we were locked inside the car. So we took the slide hammer and broke into the glove compartment. Luckily, there were the keys that open the roof and we got out. And a red light was twirling around on the dashboard. In an underground parking lot, this strobe light was going around.

Q. Are these types of alarms fairly common?

A. No, very rich people have them. The ones that have Dino Ferrari's, Turbos, expensive cars.

Q. Is the gas shutoff something affordable enough so most people can put them in their car?
A. Yes, anybody. And you will always find your car within two blocks from where it was stolen.
Q. Is this probably the best device to prevent theft?
A. I think so. It's the cheapest, and most thieves don't know about them. There's also alarms where the car is open. You put your key in, lock it, and then reopen it. So if you just go and open it, it goes off. All dealers have their license plate rims that say, for example, BMW from Century City. Take that off immediately. That tells a car thief a lot about the car. Including any with alarms as standard equipment.
Q. Like what?
A. You know what kind of stock and features are on it. By checking with the dealer, you learn the key code.
Q. What about valet parking? Is it safe?
A. Never let a person at a nice restaurant drive your car and park it for you. Always park it yourself.
Q. Why?
A. Because they duplicate your keys. I would love that job, but I'm out of the business now. But it's true, they make a little print-out of your key.
Q. Are there any other ways people can prevent their cars from being stolen?
A. Yes; by watching where you park your car. Keep your car visible. Keep it in view where other people can see it, too.
Q. What about parking lots?
A. I've been in the middle of taking a Volkswagen where I got the key on, but it wouldn't kick over. No problem, though. I had kids help me push-start the car. It wasn't even my car. There've been things on shows like *Candid Camera* where people were breaking into cars and asked people just passing by to help them. People will help. They don't even know if it's your car. Society doesn't really give a shit.

Q. If your car was stolen, where would *you* look?
A. If I had a car stolen, I would first start checking ads in the paper for parts for my year car. If I had a BMW, I would look for BMW right door, left door, or under an add that said, "have parts for BMW." Just go check it out. But I have my stuff initialed.
Q. You have initialed already so you can just go and say, "This is mine."
A. Right.
Q. Are you saying that you're safe if you own any other car besides a Porsche or Volkswagen?
A. No. Most car thieves are looking for expensive cars, Rolls Royce, for instance, where they can make some money off the parts. You're safer with a Ford Falcon, a Galaxy. People only take those for joy rides. People learn to hot-wire a car and joy ride it. Most stolen cars out here are usually found. LAPD has got a very good record for finding most stolen cars. But like I said, most are just joy riders. They're not doing it for the parts.
Q. They're not professionals.
A. Exactly.
Q. How did you first learn to get into a car?
A. Through auto shop in high school.
Q. Did they show you how to do it?
A. Well, they showed me how to hot-wire a car.
Q. Right in the high school?
A. Right in the high school. They showed me how to hot-wire it right there, but it was in case your battery was dead. But getting into a 914 or 928 (the new Porsche), where the front window is all one window and there's no wind-wing—
Q. That's the way mine is. I have no wind-wings on mine.
A. But you have rubber on your back window. I'd just take a carpet knife and cut the rubber out and pull out the whole back window. It takes twenty seconds.
Q. But that means you'd have to replace it?

A. I always have to replace something.
Q. What about the sunroof?
A. It's stupid to have on a car. Very easy access. I can take any screwdriver and pry it open. I can use a knife and get in quick. Sunroofs require even higher insurance because of the easy access. Wind-wings are easier. I just use a screwdriver.
Q. If you saw a car that had a sticker saying this car has a burglar alarm, would that deter you?
A. I won't believe it. Just a sign that says, "If you value your life, don't fuck with my car." That sign I might believe. But usually people who have alarms don't put a sign up. That's for somebody who doesn't have an alarm but wants to discourage people.
Q. What about a car that's in a garage. How safe is that?
A. Safer than it is on the street. That's what we're talking about. Where to park your car. The best place is in the garage, locked.
Q. With the garage door locked and the window into the garage locked?
A. Everything locked. The next safest place would be in your driveway. Then, street parking, and it gets worse and worse off. I used to personally take a lot of cars from shopping centers. Where there are quite a few cars.
Q. And you always have somebody else there to help you out if you need some help?
A. Exactly. I always go with a few people anyway. Somebody has to drive me. I don't leave my own car there. Usually we'd pull up next to the car we wanted. I usually don't take it, unless I see the people who get out of it, because they could return any minute. If I see the car pulling in, I know they're going to be in there for awhile.
Q. Would you enter a vehicle if there was a dog inside?
A. No.
Q. Regardless of what kind of dog it was?

A. It would depend on how badly I wanted the car.

Q. What about an airport?

A. Tell me about an airport. I had a weird experience at an airport. We went to LAX to get a Porsche. We got a ticket and pulled in through the gate. We got the car, were leaving, but the ticket wasn't stamped. The machine screwed up. My friend was driving the stolen car; we gave him the stamped ticket. When we got up there to pay for it, the lady didn't know if it was there for a day, two hours or a week. So, she took down his license number and all of the information about the car.

Q. They mark the tires with chalk.

A. Yeah, I guess they do sometimes, but she didn't know. So she wanted his identification. He gave it to her. The car was never found. We completely got rid of the car, so there was no proof that we actually took it. They still need to prove that we took it. Cars are frequently stolen from the airport. I've worked there before. If I'm going to sell a car in one county, I try to take it from another county. It's an all-night job. We usually do night work. Not too much is done in the daytime. Like I said, a car can really be gone in less than 17 seconds.

Q. Do you normally find any valuables in the cars?

A. Yes, we do. I've found cameras, video equipment, cash, checkbooks, credit cards. A lot of people leave coats. You find a lot of valuables in nice cars.

Q. Do you ever do just trunk jobs?

A. No. I've done just hubcaps and tires. I go to somebody's house and loosen all the lugs on the tires, then go back later with two people and jack the car up and take the tires off.

Q. So you have to go twice? The first time you go and do the lugs?

A. I get everything loose so it's ready.

Q. What if they drive their car away before you return?

A. They've had it. I've just jacked the car up and taken

the tires off of it, too, though, if I thought it was safe.
Q. What about the locks on tires?
A. They're easy to get off with a screwdriver. There's a little key that fits on there; put it on the socket, and all I need is a screwdriver to spin it. Those are cheap: a good whack with a screwdriver or a chisel will spin that sucker right off.
Q. Do many people in the profession use master keys or things like that?
A. Like I said, some valets use those things.

Why They Steal Your Car And How They Get Rid Of It

Q. In your profession, if it's called a profession, do you feel any risk involved at all, or are you just against society?
A. I don't feel any danger at all actually taking the car. I don't think I'd ever be caught in the act. I would run. I always leave myself an out, the same as driving. I have to plan on the unsuspected. If the thief gets caught, it's usually afterwards. I got caught four months after I sold the car.
Q. If you're in the middle of ripping off someone's car, would you have an excuse for being there?
A. No. Because I won't get caught. I'd run, I don't stand around. Leave everything, and wear gloves.
Q. Do you carry a weapon?
A. No, that's armed robbery. It's a different thing. I've never even had a close call.
Q. How much time would you get for a car theft?
A. Three years in the state penitentary.
Q. Did you know that when you did your first job?
A. No.
Q. Would it have made a difference?
A. No.
Q. What about a mandatory sentence? If you knew you were going to have to go to jail?
A. Yeah, it would. Jail is a deterrent. I'm losing sleep

over it. I've got five counts; 3 receiving and 2 stealing. But they can't charge me for receiving and stealing the same thing. I either steal it or receive it. If I steal it, I've already received it. Do you know what I mean? They charged me for stealing the car and receiving. So out of five counts, three are already kicked out. I plea bargained and got everything down to a misdemeanor which is maximum one year.
Q. Instead of fifteen?
A. Exactly.
Q. What would have deterred you from ever doing it?
A. Well, my probation officer in the report said that my only regret was that I got caught.
Q. Did any of your friends ever get caught before you?
A. Yes, all of them did. The lightest sentence right now is 16 months.
Q. That won't stop you by itself?
A. I haven't done any cars since I've been arrested.
Q. I mean before that. When all of your friends were arrested.
A. No, not really.
Q. In most situations, how many people are there involved from the time you steal the car, to the time you sell it?
A. Three or four. Some people have their car painted.
Q. You told us that there was a route you take once you get the car. How does that work?
A. Back in the beginning we were talking about buying the car. So, now I go and buy the crashed car from the person in the newspaper. He sells it to me. Now, I do the change on the numbers. Then I sell the car to the other person. So when it gets back, if it ever gets back to the person who originally crashed the car, the buyer is going to say he sold it to a guy with blond hair. The guy who actually got caught with the car thought it was somebody with black hair. They don't actually know who did the switch. That's the whole thing, the

person who buys the car doesn't sell the car because I can get caught for doing the switch myself. Two people are involved. Because if I buy it and you sell it, we're two different people and we can't be connected. There's no way the police can actually say I changed the numbers on that car.

Q. The crashed car has to be the same exact type and year as the other?

A. Not really the year. I can get 1973 paperwork and go up to 1976 on the car.

Q. As long as the car is the same?

A. Exactly. Like I said, when a policeman pulls you over, all they want is registration and license number. They'll hardly ever look at the numbers on your car.

Q. And even if they do, they're going to match.

A. Yeah, but they can tell if it's been tampered with.

Q. What about the one that you torch on the Volkswagen, underneath the seat?

A. I torch it way in the back.

Q. But they can see that it's been torched?

A. No, I do it way in the back. You gotta be good.

Q. What do you mean way in the back?

A. Well, I've got to cut it out. The numbers are only five inches long. But I put a whole piece in there.

Q. You take that out of a car and insert the entire thing? They're looking right around the number for tampering, when it's out to the side. What about the matchup of the engine with the speedometer? They could go back to the dealer and look at the service record.

A. Engines are changed day in and day out. Engine numbers are identifying parts, but mainly they're used for part numbers. The numbers that are on your engine are used to identify the engine part number, not to identify the car. You can almost put an entire stolen engine in a different car and bring it to the DMV, and there'd be no questions.

Q. Where are these jobs done?
A. I would rent out people's garages.
Q. People's garages?
A. Yeah, I told them I work on cars. We try not to keep a car in there for very long at all. Two days at the most. In and out. It doesn't really take that long. I can change the numbers on a Volkswagen in ten minutes.
Q. Is there a market available to sell the cars?
A. I put it in the newspaper and sell it to the first person who comes along with cash. I sell it as cheap as I can. If it comes up where I want $10,000, and they come up with $8,000, I sell it to them. I get it out of there.
Q. Always cash?
A. Always. Cash and carry. Usually all the cars I get into are never in my name. I sell legal cars now. I buy myself an old Volkswagen; I know where to get "warm" parts.
Q. Warm parts?
A. I just sell it to the first person who comes along. Usually, like I said, I take the car back, so there's absolutely no way possible to get caught.
Q. You just sell the car again, or use it for the parts?
A. I can also follow it right to the junk yard. Follow up who buys it after that.
Q. Aren't the police suspicious knowing there's a pattern like this?
A. Yeah, those "Bad Cats," Burglary Auto Division. They stamped that right on my medicine chest when they came into my house to arrest me. They came in and put up that stamp that says "Bad Cats, Gotcha." It's a great joke.
Q. Real funny, huh?
A. Yeah, they'll follow you. Especially because we dealt mainly in Porsches. Since we quit, Porsche thefts have gone down tremendously. Tremendously. I think I'll switch from cars to catamarans.
Q. Those sailboats that are real popular right now?

A. Yeah. They have painted numbers. They just change the number.
Q. They don't steal them back though, do they?
A. No, there's no paperwork on them. There's no identification.
Q. How would you take them?
A. Just pull up a truck and hook it up.
Q. It's on a trailer? Just hook the trailer on?
A. We've taken dirt bikes like that. Just pulled up somebody's driveway and hooked up their trailer.
Q. So you should never keep a boat or motorcycle hooked up to a trailer and ready to go?
A. Never have it ready to go. You gotta be careful. It takes an extra minute to be cautious, but it saves you a lot in the long run. We'll get you if you don't.
Q. When you say "we," how many people are you talking about?
A. Five, four of them are in jail. So now it's just me.
Q. You stole fifteen cars in your life?
A. No, that's just what I actually sold.
Q. How many would you say you took?
A. Ninety-five to a hundred.
Q. During what period?
A. Eight months. We have a four car garage in Orange County.
Q. Now your buddies are in jail?
A. Yes, all of them.
Q. You're the only one still out?
A. Yes, but not for long.

Editor's Note: Brad was sentenced to 30 days on an honor farm, and released after only 18 days on good behavior.

Buying Or Selling A Car

If you buy any motor vehicle from someone other than a known licensed dealer, use caution. And for crying out loud, before you pay for the vehicle, check the ownership with the Department of Motor Vehicles. Do you know how many people buy stolen cars without even knowing it? Thousands.

So check the ownership.

If you privately sell your vehicle, beware, too. The guy with the big grin and the cute girlfriend can be a car thief.

The same person who won't let anyone, including a friend or relative, drive his car no matter how much they beg, when selling the car, will let some grinning Joe Schmoe take it off on a "test drive." They call it that because it "tests" your stupidity.

If you forget to make sure they have a valid driver's license, you're careless. If you don't ride along on the "test drive," you're ignorant. And if you sell the car and take a personal check, instead of cash or a cashier's check, you deserve the bank memorandum attached to the check, stamped "ACCOUNT CLOSED," that you find in your mailbox several days later.

And ladies, be careful who you go off with alone in your car on that "test drive." He may seem charming, but charming has gotten a lot of women raped.

A note about that precious paper known as the "pink slip." It isn't wise to keep it in your car. It's your certificate of title, and if the car gets stolen, the thief can alter it to show him as the owner of the car. And don't sign it until you transfer ownership. If you do, your only making a thief's job easier.

What about loaning a car to someone? We've talked about buying and selling, but what about loaning?

Well, many potential clients call Nick Harris Detectives requesting that we locate their car. They had permitted a friend, an acquaintance, or a relative to use it. The car wasn't returned when promised. So the police are called and a stolen car report is requested, but the police *will not* take one, and, furthermore, won't help in recovering it. Only the aid of a private investigation agency can help them. Why won't the police?

The law points out that if you loan your car to anyone, *you* must assume the risk of the return of the vehicle, as well as any financial responsibility in the event of an accident.

If you want to be nice to your friends and relatives, don't loan them your car, unless it's an emergency. Give them cab fare or bus fare instead. You'll keep them as friends and friendly relatives much longer that way.

We all know about protecting our car from the thief, but most people never think about this ....

How To Protect Your Car While In Repair

Have you ever received your car back after being repaired to find that something of value left in your car was missing?

Have you every noticed the amount of miles driven before and after leaving your car at a repair shop?

How many times have you had work done on your car, only to find later that you could have had the same work done a lot cheaper?

Do you find yourself telling the auto service advisor at your local dealer what you think specifically is wrong with your car?

Do you always request the old parts back, and then carefully look at them?

How many times have you told the service advisor that you know little or nothing about car repairs?

There was a time not long ago, that the cost of an

automobile today would purchase a home. That was yesterday. Today, if you're the average person, your automobile is your second most valuable property.

Every automobile, sometime in its lifetime, will require repair or service. When this happens, you're vulnerable for a rip-off. Not all mechanics are crooks; most are honest. But how can you tell?

The foregoing questions should be answered when the day comes that you must leave your car for service.

Most service centers have a sign indicating that they're not responsible for your personal items in your vehicle, glove compartment and trunk.

To prevent duplication of your keys for future burglary of your premises, when you leave your vehicle, don't have your home or office keys on the key ring.

Always note your odometer reading, even if the service advisor noted it on the estimate sheet. Why? Some mechanics may use your car for pleasure purposes. If the mileage is excessive, beyond normal testing, demand an explanation.

Most mechanics receive a commission on the work they do on your car. When a surgeon suggests you have major surgery, it's usually wise to get a second opinion. So if a mechanic suggests some major repair, you may do best by getting a second or third opinion before you spend your hard-earned money. *Secure a firm "estimate" in writing.* Don't authorize any work in excess of the agreed estimate, until you have consulted another mechanic.

Don't diagnose the problem. If your car won't start, don't say, "I think I need a new battery" to the service advisor. If you do, they may just sell you one, rather than clean the battery straps or tighten the fan belt.

If you *know a little* something about auto repair, let it slip out. If you *know nothing* about it, keep quiet.

If parts are replaced, request that they give you the old parts. This gives you no absolute assurance that they came from your car, but it could keep the mechanic honest.

When you get your car back, demand an explanation for every charge. Often, "adjustments" are made if the bill is disputed.

IV
PURSE-SNATCHING, PICKPOCKETING & PARENTAL KIDNAPPING

Purse-Snatching Prevention

Swindlers, con men and thieves are ganging up on senior citizens because they're the most vulnerable. Yet you don't have to be in your golden years to fall victim.

Purse-snatching tops the list of the most frequent crimes perpetrated against the elderly. The loss of your purse's contents may cause you an inconvenience by having to replace driver's license, credit cards, etc., but worse of all, many purse-snatch victims are physically hurt. The thief, usually a teenager, or even younger, often strikes the victim from behind, knocking her down, resulting in a broken back, spine or neck.

During our interviews with experts throughout the nation, the best advice we found is don't carry a purse if you don't have to. If you must carry one, hold it under your arm. If you're right-handed, carry it under the left arm, leaving your "best" arm to fight off an attacker. If you're wearing a skirt, sew a pocket in it to carry essential items. When shopping, do so with another person. Most purse-snatchers told us they prefer that the "target" be alone, and would in most instances pass up two or more ladies, as this involves more voices yelling and screaming, and more who may want to fight back.

Easy Pickin's For The Pickpocket

Look in the Yellow Pages under "schools," and you'll find private institutions that teach you how to become a nurse, bartender, private investigator, or even an attorney. There are several unlicensed schools that will teach you how to "let your fingers do the walking," but these pickpocketing institutions aren't listed in your Yellow Pages.

How many professionally trained pickpockets there are is anybody's guess. But each day, thousands of victims are separated from their wallets.

PURSE-SNATCHING, PICKPOCKETING & PARENTAL KIDNAPPING

Can you afford to lose yours? The average person has three credit cards. By federal law, the maximum personal loss is usually $50 per card. Then there's the cash—not protected by law, and usually not insured. The inconvenience of replacing your driver's license, social security card and other important documents is time-consuming.

You can beat the pickpocket pro at his own game by not playing by his rules. First, *make it a habit not to carry a lot of cash; don't put all of your money in one place.* If you're out with your spouse, each of you should carry part of the money.

Never carry more than you can afford to lose. The trained pickpocket often observes his "mark" before attempting the score. If you expose a large sum of money, you may become the target. If you have to carry a great deal of cash, only show small bills, sufficient to pay for your immediate purchase. Keep the rest hidden in your wallet.

Thieves are out in force in crowded areas, i.e., stores, restaurants, airports, horseracing tracks, banks, to name

just a few. *Anyplace there is a crowd of people, pickpockets are there, too, so be extra alert.*

Most men carry their wallet in the back pocket of their trousers. This is the worst place to carry your money. Besides, by sitting on your wallet, you can cause a sciatic injury. *The safest place to keep your wallet is inside your jacket pocket.*

Ladies, don't keep your wallet inside a see-through handbag. The best protection against the pickpocket is a closed zipper handbag, preferrably with a shoulder strap.

The schools for pickpocket professionals teach all the tricks of the trade. They practice in the art of catching you off guard. Street magicians do something similar, although legal, which is to use diversion to conduct their magical entertainment.

So when you're in a crowd and suddenly you feel yourself jostled, *turn around and look the person or persons square in the eyes.* If the person were about to set you up as a "mark," in most cases, he would drop you and find another pigeon.

Often some pickpockets operate in pairs. Sometimes a team of three. One is the "decoy"; the decoy can be an attractive woman, a small child, or someone old enough to be your grandparent. They don't wear masks, either; they look like anyone in the crowd.

The University of Pickpockets has a lesson plan. It contains hundreds of diversions to keep your attention focused upon exactly what they want you to see. Among the most popular methods is the clumsy spilling of a drink, ice cream or anything else on the mark's clothing. Faking an accident by stumbling near you is often used. Whatever method used, it distracts your attention, putting it on the decoy, so that the other conspirator can make the move to the wallet. Your wallet. Most of the time they're out of sight by the time you've discovered your wallet has vacated its pocket.

To graduate from the University of Pickpockets, each

person has to observe a professional rule. A rule the pickpockets keep if they want to survive and stay out of prison. The rule is this:

"Take only from those who don't protect their valuables. Don't risk being caught—because there are far too many pigeons waiting to be picked and plucked."

Parental Kidnapping

Nick Harris Detectives, Inc., headquartered in Sherman Oaks, California, estimates that 150,000 children are "stolen" by one parent from the other each year. Nearly one-fifth will never see *both* parents again.

Founded in 1906, the Nick Harris Missing Persons Bureau maintains a national network specializing in tracing stolen children. Their success record is over 90%, with more than one million total assignments.

The suspect of a parental kidnapping usually involved a separated or divorced parent. The results become traumatic for all involved: the children, the mother and the father.

Why does one parent take the children from another? Unlike real kidnappers who do it for ransom, the parent involved most frequently has no criminal background.

Sometimes a parent will "steal" their children to punish their former spouse. This is a vengeful person who often doesn't really want the youngster.

Others feel that the other parent is unfit, and that for the best interest of the children, they should have them.

Nearly twice as many fathers "steal" their children. Many experts believe they do so because the mother either received child custody or they feared the court *would* give the children to the mother.

Based upon the files of Nick Harris Detectives, most stolen children usually are eight years old or less. It's not uncommon that the suspect-parent would only take one of the children. Often the father will take only the boys while the mother only the girls.

The victims—the children—are caught in the middle. They are the *football,* and the parents line up on opposite sides, and try to take possession and run.

It's not unusual for one parent to "snatch" the children from the other, hide, then be located, and have the children stolen back. There are case records indicating some children were "stolen" over ten times during a few years. A long, drawn out game of hide-and-go-seek.

Is it a crime to "steal" your own child, married or not?

The first thing most persons do when their child is stolen, and they suspect their former spouse or a natural parent was involved, is to call the police for help. Too frequently, it's the position of law enforcement that parental kidnapping or child-stealing is really a domestic matter and not a criminal one. Little or no assistance is offered to the grieving parent.

Not many years ago, a stolen child was a serious matter to the police, but today, this is so wide-spread across the nation that it doesn't demand particular attention. Two decades ago, if a minor ran away from home, a missing persons report was taken, and a police follow-up investigation was made. Today, with hundreds of thousands of run-a-ways leaving home, a traffic citation receives more police concern.

In 1980, Congress took an interest into parental kidnapping. Republican Senator, Malcolm Wallop of Wyoming, was a chief sponsor of a bill to help locate the child-stealing spouse.

In cases involving interstate child-stealing (when a child is removed from one state to another) the FBI can request a search of Social Security records. This law went into effect on July 1, 1980.

While this legislation is not the absolute solution, it's a step in the right direction. The information (identity of an employer) is never current. Normally the data will be 4 to 6 months old when received by the F.B.I.

We need, but don't have, a strict uniform law in each state dealing with child-stealing. Many, but not all, states have some type of custodial interference laws, but the crime is usually only a misdemeanor, a minor crime.

California, Ohio and Wyoming make parental kidnapping a felony crime. So far, very few parents have been convicted and sentenced as felons.

Some 40 of our 50 states adopted a Uniform Child Custody Jurisdiction Act, which recognizes another state's custody order and the right to extradition. Many legal advisors tell us extradition most often is extremely difficult, if not impossible.

Less than 1% of the known child-stealers flee to another country. When they do, it's legally out of the question to get that country to return the stolen children. The only answer is stealing them back, and this is risky, and may cause the parent trying to recover the children to be arrested in the foreign country.

The odds are greater that your child will be stolen on a Friday. The months of April and September are considered the peak child-stealing periods.

The statistics provided by California Attorneys Investigators concerning parental kidnapping indicated that children who are given to the other parent for visitation or vacation frequently are not returned.

Parents without visitation rights often "steal" the children while they're at school, frequently in the playground, or when walking to or from school.

It's a criminal act in any state to break into someone's residence and "steal" the children. While this will occasionally occur, it's rare. We know of many cases where actual physical violence was used, often upon someone only taking care of the children.

Parents prefer abducting the youngsters when they're outside the yard playing. Most parents taking the children don't consider the effect it will have on the child.

Once the children are taken from the other parent, quite frequently the abductor endeavors to brainwash the youngsters, tell them the other parent is no good, and that it's best that they live with him or her, or they're bribed with new toys.

We've seen many cases where the children are returned to the other parent, and they have a hatred toward the mother or father they once loved dearly. On the other side of the coin, it's not uncommon for a parent to build hatred into the minds of children towards a spouse who has legal visitaticn rights.

In most states, no crime is considered committed by either parent taking "custody" of their children with or without the consent of the other, *unless* the court has given custody jurisdiction to one of the parents.

If you have custody, and your former spouse was ordered to give support payments, the Parents Locator Service, through your local welfare office, may be of some help if the children were stolen.

Locating the whereabouts of someone, particularly if they're trying to conceal themselves, can easily become a needle-in-the-haystack problem without the assistance of a professional private investigation agency.

Harold Miltsch was a victim of parental kidnapping. He retained the Nick Harris Detective Bureau to assist in locating his stepchild. He later became one of the most concerned activists in the nation helping other victims. He founded *Stop Parental Kidnapping* based in Rochester, New York. His newsletter, "Return Our Children" is distributed to schools, pediatricians and others, and has helped many persons relocate their stolen children.

Another former victim, Arnold Miller, who spent five years searching for his child, founded *Children's Rights, Inc.,* based in Washington, D.C, and has nearly 100 chapters throughout the nation. They offer a hotline, and provide support to victimized parents, lobby for strong laws, and publish a quarterly newspaper.

Your valuables can be replaced; children can't. One way to protect your mental stability from the vicious beating given by the loss of children by parental kidnapping is to leave open the parental tie between you and your spouse, no matter how hateful, argumentative and difficult your relationship is with this individual, because parenthood will keep you bound for the rest of your life. So you should make every effort to discuss and agree and plan your children's futures accordingly.

V
FORGERY, COUNTERFEITING, SHARKING & INDUSTRIAL ESPIONAGE

How To Write your Checks

A team of counterfeiters specializing in bogus ten-dollar bills, decided to move their operation to the Ozarks. They planned on ripping off the hillbillies.

The leader came up with a brainstorm. "Let's print twelve-dollar bills and come home with a bigger profit."

After the first batch was dry, they decided to pass the counterfeit cash in the first hillbilly bank they saw. Inside, the leader asked, "Could you give me change, please?"

The teller bared a beaming smile, saying, "Sure can! You want four three-dollar bills or two sixes?"

* * *

The counterfeiter doesn't have to go to the hills to find his prey. Nor does his "brother" the check forger. Check forgers steal millions of dollars each year from individuals and businesses. Ever wonder how much you've contributed? Well, many of us have been victims and didn't even know it.

Many years ago, we learned our lesson and never forgot it. We were in the habit of bringing the mail to the office and depositing it in the mail sack provided by the post office. In the past, we never examined each cancelled check when they returned from the bank with our statement.

To stay one step ahead of the forger, you have to know how he operates.

Well, we received notices that our insurance premiums were not paid, our credit cards were unpaid, and our phone payment had not been received yet. We knew they'd all been paid. We pulled out the cancelled checks. We were stunned.

The names of the payees were changed—very crudely—and all the amounts were raised. It took a lot of talking to convince our former bank that they were forgeries—very bad forgeries.

Later, we talked with the Postal Inspector and were told many forgers go into office building with open mail sacks, take the mail and alter the payee's name and amount on the checks they find.

Unless you've worked in a bank, you probably don't know that many banks, as a matter of policy for keeping down overhead, DO NOT verify their depositors' signatures on checks they cash, unless the check is over $1,000. Shocked? We were. In other words, if you signed your name, Santa Claus, or George Washington, or even Jesus Christ, most likely your bank would honor your check.

To start with, when you order your printed checks, don't have your name imprinted the same as you sign it. If your signature is Mary Jane Smith, only have M.J. Smith printed on your checks, otherwise, if the forger gets ahold of your blank checks, he would know how you signed it, and you'd have a harder time convincing your bank that the checks were forgeries.

Don't keep cancelled checks around, either, or someone may have a true specimen from which to forge your name on one or more of your own checks; again, it would make it difficult for you to convince the bank that it was a forgery. And if you do find yourself in this predicament, most good forgeries, with a signature only, *can't* be absolutely verified

FORGERY, COUNTERFEITING, SHARKING & INDUSTRIAL ESPIONAGE

by questioned document experts.

When you write a check, remove it from the checkbook FIRST. Why? Well, if you have a heavy hand (and most of us do when we are forking out money for the bills) and use a hard-point pen, the impression will go through and onto the next blank check, and can easily be traced. Your own "authentic" signature.

Leave no space on a check blank. If you write a check payable to an "ABC Company," and you don't draw a line in the space that follows, anyone can add OR CASH —and the check would be paid to anyone.

Get out of the habit of writing only part of the company name or using its initials. If you wrote a check to RALPHS MARKET but only put RALPHS, a forger can simply change it to read, RALPHSMITH —or any other name. A check made payable to G.M.A.C. can easily be changed to read, G.M.ACON, or any other name.

Unless you want to pay more than you agreed to, don't leave any space between the $ on the check and the first figure of the amount you wish the check to pay. So when you have intended to write $3.70, a forger changes it and cashes it for $93.70 by inserting a nine in front of the three. Forgers can alter the written amounts just as easily.

We highly recommend a check protector. It's a worthwhile investment for all businesses and many individuals as well. Keep the checkwriter inked well, because a weak impression will permit the forger to use his own checkwriter fully inked, thus, overprinting the amount you entered.

And we can't stress the importance in examining each check returned by your bank with your monthly statement. Make certain the payee's name has not been changed, the amount has not been altered, and the check was actually signed by *you* —not Santa Claus or Jesus Christ.

Your Money Cards

Let's say, you're wiser than some. You proudly tell your friends, "I never carry large sums of money. Just credit cards and my checkbook."

You may not think you're carrying a lot of money, but even the middle-class, blue-collar worker, with a MasterCard or Visa credit card, holds $500 or more in his/her wallet. Add a few thousand dollars more for the oil company cards, the Sears charge card, and others, and you'll find that the average American today is walking around perhaps with more than they have in their savings account. If you have good credit and an executive's income, you may also hold an American Express card. It'll buy a dozen of the finest custom made suits, at a flash of the card, a five-carat diamond ring, and, who knows, maybe even a Lear jet. Do you still think you don't keep large sums of money on you?

Today there is a lucrative business for credit card sharks. Using fraudulent schemes, they rip off both you and the credit card companies. The credit card companies lose nearly $500,000,000 each year. And with 675,000,000 credit cards in circulation, often the loser is the credit card holder—you.

Here are some of the ways they reach into your wallet and take your money:

You're out on the town, enjoying a good dinner, and finish your last sip of Cognac, and then you nod at the waiter, and he presents you with a check. Out comes your wallet, and you lay the "white plastic card" on the tray. Minutes later, he returns, with your card's imprint on the voucher. You add a tip, sign your name, and the waiter thanks you as you leave the establishment.

You leave the waiter $7.50 tip, and your tab was $68.40.

Your waiter tells himself, "That cheap #$&*%@, great tip." To 'make things right,' he adds a one to the tip, and when you get your bill from the credit card company, it will read $87.50, which includes the *$17.50* tip.

Later on, you may have a hard time convincing the credit card firm that you only left $7.50 tip. If you keep your copy of the receipt, that should be evidence enough. It's easy to forget the amount of the bill unless you do keep your receipt and check it, because most credit card firms don't mail back copies of the voucher.

For additional protection, make all the totals on the voucher in your own handwriting including the amount of the tab, together with the tips, and when you sign your name, after the signature, add the total amount, and draw a circle around it. Now. If service was so bad you didn't feel the waiter should receive a tip of any amount, don't leave the tip area blank, fill it in with the word "NONE."

And don't think for a moment that although the service was very good and you gave the waiter, let's say, $17.50 tip, that he has no reason to "make things right"; you still may be a subject to a rip-off.

How? This is not commonplace, but some waiters or perhaps cashiers write down your name, credit card number and date of expiration for "future use."

One can buy almost anything by mail order or telephone. All you need is your trusty credit card. They only require three things: your name, credit card number, and date of expiration. Sometimes, they even verify it: "John Smith, #5983092, expires 5-83, amount $87.30. He's good for it; your verification number is 724." Bingo! Someone, using your name, number and expiration date, ordered $87.30 in merchandise, without a signature. And guess who's going to get the bill? You. By the time you receive your bill, which could be up to 30 days after you dined at the restaurant, you may find that all of your line of credit has been used up.

You may not be legally responsible. But you're going to have to convince the credit card company you didn't make those charges. Your time is money. When you deal with a bank, concerning your personal account, they'll usually challenge you with personal information that only you would know; for instance, your mother's maiden name. But with a credit card it's just a name and number. So keep your credit card information confidential. Or start to pay in cash. But that's not the American way.

Any merchant who takes your credit card can be dishonest; most are not, some are. This time, you're shopping at the liquor store. Your purchase amounts to $53.82. The store's limit, without verification of the card, is $50. The clerk excuses himself, telling you, "I have to clear this." He goes to an area out of your view. Less than one minute passes. He returns, saying, "The card is as good as gold." And off you go with your purchases. Within a few weeks, you receive your credit card statement. Your liquor purchases were $39.73, $53.82, $28.56, $14.05, and $48.41.

You scream, "They made a mistake!" You spent only $53.82. So you go back to the liquor store for an explanation.

The manager pulls out the credit card voucher. Each amount is verified, but he notices that the signature is different on all items except for the $53.82 that was verified. You ask to see the clerk, and you're told that he no longer works there.

So, how did all of this happen?

Well, when the dishonest clerk left your sight to verify your card, he ran off four additional vouchers, and later filled them in, and removed that amount from the cash register, and replaced it with your credit charges.

How can this be prevented? Simple. Don't allow the clerk to take your card out of view.

One last credit card scheme that we feel you should know about. Let's say you're at the record store, buying a few LPs. You pay for them with the good old "plastic," but the store is busy and you get back your card, put it in your wallet without looking at it, and later find that you received someone else's credit card. Same card company, but it doesn't have your name on it. The dishonest employee will use it for a couple of days, a buying spree no less, until you realize it was taken from you. Then the thief will give your card to the next customer, and so on.

No, don't. Put those scissors down. It's not the American way.

The Loan Sharks

Today legitimate interest rates are extremely high. Buy a house for $100,000, take the standard 30 years to make the payments, and by the last year you would have paid back at least $300,000.

Three times the amount you borrowed sounds unreal, but remember that was for a period of three decades.

Let's suppose you need $500 to buy something special, or pay an unexpected medical bill, whatever the reason, you don't need to have good credit to deal with the loan shark. He has plenty of money to loan. No papers to sign, no credit bureau checks, no liens on your property, no co-signers, your employer is not informed, not even your spouse. The deal is made with a handshake.

"Need $500, no problem. Take all the time you need to pay it back. Interest is only $50 a week."

That's only 10% per week. Well worth the convenience of an easy loan, or is it? The loan shark doesn't have the guarantee the bank does. Sometimes the borrower will default. If this happens there's no lawsuit, attachment, garnishment, or repossession. If you can't pay, your "creditor" will probably write it off (not from his taxes) from his books.

No, you don't escape payment all that easy. You pay. You must pay, otherwise the other "debtors" would forfeit on their payments. How do you pay? It depends upon how much you owe, for how long, and how much of an example must be made of you. If luck is with you, it may only be a broken hand or a few ribs. If you aren't so lucky, it may be your life.

That $100,000 you borrowed from the mortgage company on your home cost you $300,000 thirty years later. If you made a similar loan from the "shark" and you had the same amount of time to pay it back, at his 10% a week, it would have cost you *seven million, eight hundred thousand dollars!*

Loan sharks have no regulation governing what they charge; it can be more or less than illustrated above. It's not uncommon to charge as much as $2,500 a year, for a $500 loan!

Your "loan" with the *shark* cannot be enforced in a court of law: loan sharking, as you know, is illegal. If you borrowed from the loan shark, and want to see him prosecuted, contact the office of the Attorney General in your state.

Just remember two things:
1) Never borrow more than you can repay, and
2) Don't ever pay more than the legal interest rate.

How They Steal Your Company's Secrets

If a burglar breaks into your office and steals your typewriter, the loss is a few hundred dollars. If he steals only the ribbon in your typewriter, and that was his target, it can cost you thousands of dollars, perhaps financial ruin.

Industrial spying is commonplace today. Your competitor would like to get his hands on your trade secrets, your customer lists, advance knowledge of the amount of your sealed bid, copies of your correspondence, and other valuable information.

Discovering your company's trade secrets is an easy task, unless you prevent it from happening. Your typewriter ribbon can be a warehouse of information if it is a common "one time" carbon ribbon found in IBM and other popular typewriters. The ribbon can be like a clandestine "bug" recording everything you put in writing.

Consider the financial impact to your company if your competitors had access to a copy of every item typed on your typewriter. Every stroke typed becomes a permanent record, just waiting to be read by the opposition. Unlike cloth ribbons, the carbon ribbons, which produce more superior work, can only be used once, and the impressions remain visibly clear.

Most secretaries throw their typewriter ribbon in the wastebasket when it reaches the end. From there, the cleaning crew disposes the contents into trash bins, which is later collected by the garbage personnel. Some high paid "garbage men" known as espionage agents, retrieve your typewriter tapes, and deliver them to your competitor.

The solution?

Instruct your secretarial staff to place finished ribbons in a designated box, which is kept in a safe, or locked location. They should be removed and burned periodically. If highly sensitive correspondence is typed, the ribbon should be taken and concealed overnight.

Carbon paper and second sheets are still in use today. They often wind up in the trash basket. So do originals with errors. Remember the high paid garbage man! He sifts through everything. Experienced secretaries use a back-up sheet, a plain piece of paper placed behind the original to protect the typewriter's platten and to produce a better original. The back-up is generally thrown in the trash, and the espionage agent, with the use of a pencil, traces the impressions.

The solution? An electric wastebasket that shreds sensitive material.

Carbonless paper, NCR or IBM (no carbon required) forms are used in offices. They come in sets, from two to five sheets, each a different color. The extra, or unneeded copy, goes into the trash containing everything typed on the original. That copy must be destroyed, or shredded.

In the 1980's, the secretary who takes dictation by shorthand is a rare find, but many are still around. They carry a steno pad. Whatever you dictate is written in Gregg, Speed Writing, or other forms of a foreign language—at least to some of us. Every word is written down. "... we will pay $157,000 for each gidget ... strike that to say, we will pay a *fair price* for each gidget ...." When the steno book is used, and that doesn't take long, it winds up in the wastebasket. After work, it should be locked and hidden. Its final resting place isn't the trash, but the shredder.

But let's say you don't have to fear the espionage agent; you don't have a stenographer. Like millions of others, you dictate all your important correspondence into a dictaphone or other transcriber. When you finish the tape, you record over it, and it's never thrown away, it's used over and over again. You're safe you think, or are you?

What about the unerased correspondence still on the tape? Your last important dictation. You left it inside the dictaphone. Hours after the office closed, the cleaning crew comes in, followed soon after by a stranger. "Hi, forgot something, just take a second," says the Espionage Man. The janitor doesn't understand what's going on and probably doesn't care, and doesn't stop him either. Mr. Espionage Man just took the tape cassette. It's better than shorthand notes because it doesn't have to be transcribed and will take less time than reading the messy typewriter ribbon; best of all, though, the tape is in your own words, and the tone of your voice can reflect your true thoughts.

Entry into your office can occur many ways. A ten- or twenty-dollar bill given to the cleaning crew is one way. Burglary, "Watergate"-style is another. Going through your outside trash bins is the easiest and safest method. Few people are arrested for rifling through someone's trash.

Keep your secrets secret. Buy a shredder and use it consistently. Orientate your secretarial staff to be conscious of copies, ribbons, cassettes, and other sensitive data. If you follow these simple guidelines, you may never find yourself being outbid, or in other ways, ripped off. Being at "a loss for words" is one thing, but when that loss can cost you money, it could mean financial ruin.

CRIME: THE SOLUTION

As long as the criminal believes that *crime pays,* crime will increase. When nearly 75% of all persons arrested went to trial on criminal charges, and weren't convicted, we must wonder who the law favors, the criminal or the victim?

Some were probably innocent. The law is clear: everyone is innocent until proven guilty. But what about the vast number who were actually guilty, and escaped through legal loopholes and are free to continue their life of uninterrupted crime?

They'll return to the streets and murder, rob, rape and victimize innocent persons. Today, crime in the United States has reached epidemic proportions. We're no longer safe at home, at work or in the streets. What should be done? Here are two common viewpoints:

The Conservative View:
Criminals belong behind bars... lock them up and throw the key away.

The Liberal View:
Criminals deserve another chance ... rehabilitate them, and return them to society.

George Putnam is a noted Los Angeles television newscaster and radio talk-show host, a conservative, and as

American as Howard Johnson's. During a recent lunch, George told us about a T.V. debate he had with the then-liberal comedian Mort Sahl. Their former television show was very popular, as the two personalities shared opposite views. Except for the day Mort told George, "A conservative is a liberal who's been mugged."

Regardless of the stand you take, liberal, moderate, conservative, or undecided, you should be concerned about crime. Throughout this book we reported statistics, thousands of them, and each statistic represented a person, a victim. Violent crimes have been committed against persons close to both authors. Until crime personally touches upon your life, your loved ones', friends' or neighbors' lives, you may not cry out for justice, and demand punishment.

Contrary to public belief, most hardened criminals aren't locked away in institutions: they walk the same streets each day as we do. Among the most noted now incarcerated are Charles Manson and Sirhan Sirhan, who one day will be set free.

Are the scales of justice really balanced? Manson and Sirhan will be young men when they're released. But what about "Uncle John" Davis, as his fellow inmates call him? They locked this man up in 1922 and "threw the keys away." On September 11, 1981, John Davis celebrated his *105th* birthday behind bars at the maximum security Central Correctional Institution at Columbia, South Carolina.

He didn't mastermind the senseless killings of innocent persons. He didn't kill a United States Senator. He was convicted of stealing $5 and a used watch. No one was physically injured during this "crime." Back then, burglary was a capital crime in South Carolina. Punishment usually included hard labor on the chain gang.

John Davis was not a real criminal as we know today. His motive was different. He thought he was a victim of injustice, and took the law into his own hands. For fifty-nine years, the thoughts of his encounter with the law ran through his mind. It began the day he ordered—and paid for—a tailor-made suit. It didn't fit. The merchant reportedly refused to alter the suit. There was no small claims court in those days, and Uncle John decided to get his money back. He entered what he thought was the merchant's residence and took only $5 and a watch, about equal value to the price he paid for the misfitted suit. John made a mistake. He burglarized the wrong house. Of course, even if he had committed the burglary at the merchant's residence, by law, he was committing a crime. Today, the sentence would have been probation, and not what John got: more than half a century behind bars.

Uncle John is the highest ranking trustee in this prison and has many privileges. When he wants, he can leave as a free man (he has more than paid his debt to society), but this 105 year old man, without family, without friends on the outside, without anyone to keep and feed him, has chosen to spend his final days in prison.

There are thousands of hardened criminals who repeatedly commit violent crimes and who are also behind bars, but want to cancel their debts to society before their time.

Some criminals set themselves free. One out of every forty-three prisoners makes an escape attempt. Six percent are successful and return to the streets and a life of crime. One prisoner who attempted escape was James Earl Ray, the convicted assassin of Dr. Martin Luther King. He was recaptured, and, in 1981, fellow inmates attempted to assassinate *him*. He survived multiple knife wounds.

The number of criminals now incarcerated would be equal to the population of the city of Tucson, Arizona or Albuquerque, New Mexico. Surprisingly, there are more convicted criminals now in prison hailing from the nation's capital, the District of Columbia, than from any state in America. D.C.'s number is five times greater than the total body of Congress and the Senate.

More criminals hail from Washington, D.C., than California, New York or Illinois. The second greatest number of prisoners come from Georgia, followed by North Carolina, South Carolina, and Delaware.

In some not-so-civilized nations, there are no jails or prisons. Why? Punishment. If you steal something, they cut off your finger for the first offense, and your hand for the second offense. If you commit rape, they cut your testicles off. This is not the solution to ending crime in our "civilized" United States, but it's food for thought. Crime in those "backward" nations is almost unheard of.

The purpose of this book was to make you aware of crime as it really is, and to prepare you to prevent it from happening. This alone is not enough. Your vote is also part of the combination to crime prevention. Elect anti-crime candidates with proven records next time you put your (X) for President, Senator, Congressman, State Attorney General, District Attorney, Sheriff, Judge, or any office where crime prevention is under their control.

You don't have to wait until the next election to do your part to help stop crime. Write to your elected officials, state and federal, and demand that they pass new laws to make your streets safe again.

In 1977 co-author Milo Speriglio was a candidate for one of the nation's highest public office—Mayor of Los Angeles. His platform was Law and Order. *The Los Angeles Times* reported, "Speriglio has come out with the most sweeping program of all the mayoral candidates to combat crime." Crime was not an issue then. Tom Bradley was re-elected. Today, crime is the number one issue on the minds of Americans.

* * *

CRIME: THE SOLUTION

Crime can be decreased if stronger laws are enacted and enforced. It will never stop altogether. Here's our concise program to combat crime:
1. *Every state in the nation must have a death penalty.* More important, it must be implemented. The U.S. Supreme Court in 1966 decreed individual states may enforce the death penalty. Only 30 of the 50 states have yet enacted capital punishment statutes. At this writing 753 criminals have been sentenced to die. Only four have been executed. The death penalty won't be a deterrent until the criminal receives that ultimate solution!

Newsweek Magazine recently conducted a national poll, and determined two-thirds of Americans are in favor of the death penalty. Those against it often cite from the Bible, "Thou shall not *kill.*" Many people are unaware that the Scripture was translated more precisely from Greek. The interpretation actually reads, "Thou shall not commit *murder.*"

Justice must be swift. We believe the death penalty should be carried out no longer than one year from the date of sentencing. All appeals could be completed within this period. In the past there have been cases where criminals waited for years on death row until executed; most were never executed at all. One of the most noted was Carol Chessman, the notorious "Red Light Bandit." He lived 12 years in the shadow of death. The rapist-kidnapper was executed May 2, 1960, when at 10:00 A.M. the pellets dropped in the San Quentin gas chamber. We endorse the death penalty and the quick execution of it, for crimes committed for which society demands this penalty.
2. *The penalty must fit the crime.* The penalty must be severe if it's to be a deterrent. Our objective is to prevent as many crimes as possible from happening. If the criminal knows before he commits a crime that the punishment will be sure and severe, he may think

twice about committing it.

One of the best deterrents is mandatory sentencing, a minimum time that must be spent in jail without probation. So-called "first offenders" usually receive probation. The judge or jury frequently looks upon the convicted person as one who made a mistake, and should be given a second chance. In 90% or more of the crimes, this was not the first mistake, it was the first time they were caught. For the few first offenders, the mandatory sentence will probably end the beginning of a life of crime.

Life in prison, as we know it today, usually means the offender can be parolled within seven years of the sentence. In *our* program, life in prison means to stay in prison for the rest of the criminal's natural life. Under our proposal, the convicted criminal will have to serve at least the minimum time in prison. Eligibility for parole can come anytime thereafter. The maximum sentence for repeaters should be based upon the severity of the crime.

The * denotes a victim seriously injured:

CRIME: THE SOLUTION

CRIME	PUNISHMENT	
	Minimum	Maximum
MURDER		
Premeditated 1st degree	30 years to life	Death
2nd degree during the commission of any crime	20 years to life	Death
2nd degree not during the commission of a crime	7 to 15	Life
FORCIBLE RAPE	12 to 20	Life
Committed upon a person 12 or under or 65 or older, or handicapped	25 to Life	Death
AIRPLANE HIJACK	25 to 35	Life
*	Life	Death
ARMED ROBBERY	10 to 20	Life
*	15 to 35	Life
BURGLARY	5 to 9	15 years
*	10 to 15	Life
PURSE-SNATCH	3 to 5	8 years
*	6 to 10	20 years
VEHICLE THEFT	2 to 4	7 years
*	6 to 10	20 years
MUGGING-ASSAULT	6 to 10	20 years
KIDNAPPING	15 to 25	Life
*	35 to Life	Death
ARSON	5 to 10	20 years
*	10 to 25	Life
CHILD MOLESTING	20 to Life	Death

In our program, punishment is extremely severe for those crimes which cause physical injury to the victims and for crimes most often repeated.
3. *Plea bargaining must end.* This is a "deal" between the criminal and the prosecutor. For example, the suspect is arrested for burglary; he agrees to plead guilty for simple trespass, which carries a light sentence, and the prosecutor accepts this rather than prosecuting the case in court. The prosecutor wins more cases, but the criminal is really the winner in bargaining for a plea. Prosecutors will justify the practice by pointing out the savings to taxpayers in court cases, and in speeding up trials, lightening their case loads.
4. *The "Use a Gun Go To Jail" theory.* Our platform is that all convicted criminals go to jail for a mandatory term as outlined above. We have added this: if any weapon, gun, knife, etc., is used in the commission of a crime, the minimum sentence be extended by no less than 10%. In addition, any crime where an on-duty law enforcement officer is physically injured, an additional three year minimum sentence shall be imposed.
5. *Juveniles, from age 12 up, must be treated as adults and given stiffer punishment in juvenile court.* Our investigation disclosed that the majority of all crimes are committed by young persons. Juveniles who commit violent crimes go virtually unpunished. They become the hardened criminals of the next generation.
6. Violent crimes committed against children, the elderly or the handicaped, who are usually not capable of defending themselves, should carry *additional compulsory time in prison.*
7. Our final recommendation concerns *compensation to the victims of crimes.* At the time of the criminal trial, if the accused is found guilty, the judge or jury should render a civil verdict to monetarily award the victims for all *medical costs, pain and suffering, and*

punitive (punishment) *damages*.

If the criminal has personal assets to cover all or part of the civil award, they should then be turned over to the victims or their heirs. When no assets or only partial assets are owned by the defendant, then while in prison, the criminal shall be required to work and be paid a reasonable salary of no less than one-fourth the current minimum wage. From the amount paid, two-thirds should go toward the award for victim compensation, and one-third returned to the state to pay the cost for incarceration. If the convict refuses to work, then his privileges should be taken away.

* * *

Our program is very strong. Perhaps society is not ready to accept it. If it were in progress right now, we believe one-third of all crimes would stop. Until such a day comes when the criminal is afraid of the crime and punishment, your only protection will be self-protection.

When law-abiding persons watch criminals escape their due punishment, because the state is unwilling or unable to carry it out, this can lead to anarchy, which pretty much exists in such things as the punk rock culture, and, down the trail, may lead to the return of vigilante groups. None of us want this; all we want is to be safe from the criminal.

How to Protect Your Life and Property has shown you the way to daily survival. We made you aware of crime. We showed you how to prevent it from happening to you, and what to do if you do become a victim. Now, it's up to you to survive—to protect your life and property.

EDITOR'S NOTE

After this book was set in type an initiative measure was circulated to California voters. Called The Victim's Bill of Rights, it could turn the tide against hardened repeat criminals. This Bill covers much of the crime prevention solutions offered by your authors. The petition must be signed by 553,790 registered voters to be placed on the ballot. The voters will then decide if the State Constitution will be amended.

The Citizen's Committee to Stop Crime is behind the initiative. Their chief spokesperson, George Nicholson, Senior Assistant Attorney General said, "I have just received a very disturbing report on crime ... the situation is extremely grave. There is roughly a 50% chance that you or a close friend or family member will be the victim of a serious crime, like murder, rape, armed robbery or burglary, within the next 14 months.

"You can't move away from it. You can't escape it. More importantly, you can't ignore it." Nicholson added, "Violent crime is completely out of control ... the habitual offender is turned loose to commit more violent crimes even before his last victim is out of the hospital. It hasn't always been that way."

The eyes of the nation will be on California. Just as with Proposition 13, the Golden State started the movement that was carried all the way to the Nation's Capitol. Will the voters exercise their rights, and restore common sense and decency to the criminal justice system by placing the VICTIMS' rights before the criminals' rights?